Traffic and Mobility

Springer
Berlin
Heidelberg
New York
Barcelona
Hong Kong
London
Milan
Paris
Singapore
Tokyo

Werner Brilon
Felix Huber
Michael Schreckenberg
Henning Wallentowitz
(Eds.)

Traffic and Mobility

Simulation – Economics – Environment

With 136 Figures and 22 Tables

Springer

Editors

Werner Brilon
Lehrstuhl für Verkehrswesen
Ruhr-Universität Bochum
D-44780 Bochum, Germany
e-mail: Werner.Brilon@ruhr-uni-bochum.de

Felix Huber
Lehr- und Forschungsgebiet Umweltverträgliche
Infrastrukturplanung, Stadtbauwesen
Bergische Universität-Gesamthochschule Wuppertal
Pauluskirchstraße 7
D-42285 Wuppertal, Germany
e-mail: huber@uni-wuppertal.de

Michael Schreckenberg
Physik von Transport und Verkehr
Gerhard-Mercator-Universität Duisburg
Lotharstraße 1
D-47048 Duisburg, Germany
e-mail: schreck@traffic.uni-duisburg.de

Henning Wallentowitz
Institut für Kraftfahrwesen Aachen
Steinbachstraße 10
D-52074 Aachen, Germany
e-mail: wallentowitz@ika.rwth-aachen.de

Mathematics Subject Classification (1991): 82-XX, 35Qxx, 65Cxx

Cataloging-in-Publication Data applied for

Die Deutsche Bibliothek - CIP-Einheitsaufnahme

Traffic and mobility: simulation - economics - environment; with 22 tables / Werner Brilon... -
Berlin; Heidelberg; New York; Barcelona; Hong Kong; London; Milan; Paris; Singapore; Tokyo: Springer, 1999
ISBN 3-540-66295-2

ISBN 3-540-66295-2 Springer-Verlag Berlin Heidelberg New York

Cover design: *design & production* GmbH, Heidelberg
Typesetting: Camera-ready copy produced from the authors' output file

SPIN 10723927 46/3143/lk - 5 4 3 2 1 0 - Printed on acid-free paper

Foreword

Anyone who reflects on the future of society cannot do so without at the same time thinking about the future of our transportation systems. The dilemma is obvious. On the one hand, mobility must be maintained as it is crucial to economic development and because people are eager for individual mobility. On the other hand, traffic imposes heavy burdens on people and on the environment, on cities and communities and on our national economies. Finding a solution to that dilemma seems to be difficult, in fact we have not even developed a rough idea of how it could look like.

This is why the North Rhine-Westphalia Science and Research Ministry came up with the plan to work out a well-founded scientific basis on which to solve the problems inherent in our transport system. A research network has been established and sponsored with government funds for a period of three years with a view to realising that objective.

The "Traffic Simulation and Environmental Impact" research network is composed of researchers who have an excellent reputation as North Rhine-Westphalia traffic experts. Cutting across various disciplines of knowledge, the network aims to integrate transportation and natural sciences, particularly physics and mathematics, in a move to profit by the synergy between technical know-how and innovative methodology.

The present volume is intended as a progress report and a prologue to the forthcoming international colloquium which represents the highlight and at the same time the end of the three-year project funding period.

My impression is that much has already been achieved, especially as far as road traffic simulation is concerned. The models developed by the network help understand and simulate swiftly and realistically a great number of phenomena such as the build-up of traffic congestions. The models have attracted much attention at the international level. This goes also for the network's models for the dispersion of pollutants where the experts succeeded in combining traffic with pollutant dispersal models.

It goes without saying that, despite the results that have been achieved so far, much remains to be done to enable us to fully grasp and influence the transport system in its entirety. Optimising traffic is also concerned with forecasting, and it is desirable that we be able to make reliable prognoses on road traffic. Concepts of managing and regulating tomorrow's traffic need to be devised and made available to the relevant authorities as a basis for decision-making and planning.

Moreover, it will be vital to integrate the various transportation sub-systems over the medium term, which is a task of outstanding societal importance given that it involves nothing less than providing transport alternatives and even avoiding traffic. During the three-year project funding stage which now comes to a close the foundations have been laid on which new initiatives and practical applications can be built.

I would like to express my thanks to the members of the research network for their commitment and fruitful work, and I wish the conference every success.

Gabriele Behler

North Rhine-Westphalia
Minister of Schools, Education,
Science and Research

Preface

This book contains elaborated manuscripts of the presentations given at the international workshop "Traffic and Mobility: Simulation – Economics – Environment", organised by the Institut für Kraftfahrwesen (ika) at the RWTH Aachen. The aim was to inform about the latest developments in the field of traffic simulation and to serve as an interdisciplinary forum for mutual exchange of traffic practitioners and theoreticians all over the world.

At the same time this workshop was initiated as a final presentation of the results, achieved by the North-Rhine Westphalian (NRW) research cooperative "Traffic Simulation and Environmental Impact" NRW-FVU (Verkehrssimulation und Umweltwirkungen) founded by the Ministry of Schools, Education, Science, and Research in NRW.

The idea was to present not only the results of a 3-year research cooperation consisting of twelve scientific institutions from NRW, but also from internationally acknowledged experts in this field. The subjects range from traffic flow theory over economic factors of traffic generation to environmental effects. The success of the workshop encouraged us to publish the results as a proceedings volume, which will now be accessible to a broad audience interested in the topics treated.

We are very grateful for the research support by the Ministry of Schools, Education, Science, and Research (MSWWF) of North-Rhine Westphalia and for the financial sponsoring of the workshop by Siemens AG and Forschungsgesellschaft für Kraftfahrwesen Aachen (fka). We would like to thank V. Jentsch for his useful advice, the Institut für Stadtbauwesen Aachen (ISB) for its active support, L. Kang-Schmitz and D. Neunzig for their local organisation, B. Dahm-Courths and L. Neubert for editing these proceedings and the advertising agency **z.B.** for the cover picture.

Duisburg, July 1999

W. Brilon
F. Huber
M. Schreckenberg
H. Wallentowitz

Contents

List of Participants

1. T.A. Arentze, Eindhoven University of Technology, Urban Planning Group, P.O. Box 513, Mail Station 20, 5600 MB Eindhoven, The Netherlands, t.a.arentze@bwk.tue.nl

2. K.W. Axhausen, Institut für Straßenbau und Verkehrsplanung, Leopold-Franzens-Universität, Technikerstr. 13, 6020 Innsbruck, Austria, k.w.axhausen@uibk.ac.at

3. J. Barceló, Laboratori de Simulació i Investigació Operativa, Departament d'Estadística i Investigació, Operativa, Universitat Politècnica de Catalunya, Pau Gargallo 5, 08028 Barcelona, Spain, lios@eio.upc.es

4. R. Barlovic, Physik von Transport und Verkehr, Gerhard-Mercator-Universität Duisburg, Lotharstr. 1, 47048 Duisburg, Germany, barlovic@traffic.uni-duisburg.de

5. H. Baum, Institut für Verkehrswissenschaft, Universität zu Köln, Universitätsstraße 22, 50923 Köln, Germany

6. K.J. Beckmann, Institut für Stadtbauwesen, RWTH Aachen, Mies-van-der-Rohe-Straße 1, 52074 Aachen, Germany, k.j.beckmann@isb.rwth-aachen.de

7. M.G.H. Bell, Transport Operations Research Group, Dept. of Civil Engineering, University of Newcastle, Claremont Tower, NEI 7RU, Newcastle upon Tyne, UK

8. R. Böning, Zentrum für Paralleles Rechnen (ZPR), Weyertal 80, 50931 Köln, Germany, boening@zpr.uni-koeln.de

9. W. Brilon, Lehrstuhl für Verkehrswesen, Ruhr-Universität Bochum, 44780 Bochum, Germany, Werner.Brilon@ruhr-uni-bochum.de

10. W. Brücher, Institut für Geophysik und Meteorologie, Bereich Meteorologie, Universität zu Köln, Kerpener Str. 13, 50937 Köln, Germany, bruecher@meteo.uni-koeln.de

11. J. Casas, TSS-Transport Simulation Systems, Tarragona 110-114, 08015 Barcelona, Spain, info@tss-bcn.com

12. C.F. Daganzo, Department of Civil and Environmental Engineering University of California, Berkeley, CA 94720, USA, daganzo@ce.berkeley.edu

13. S.T. Doherty, Department of Urban and Regional Planning, University of Laval, Quebec City, Quebec, Canada, G1K 7P4, doherty@grimes.ulaval.ca

14. A. Ebel, Institut für Geophysik und Meteorologie, Projekt EURAD, Universität zu Köln, Aachener Str. 201-209, 50931 Köln, Germany, eb@eurad.uni-koeln.de

15. G. Eisenbeiß, Deutsches Zentrum für Luft- und Raumfahrt (DLR), Linder Höhe, 51147 Köln, Germany, gerd.eisenbeiss@dlr.de

16. J. Esser, Los Alamos National Laboratory, MS M997, Los Alamos, NM 87545, USA, esser@santafe.edu

17. K. Esser, Institut für Verkehrswissenschaft, Universität zu Köln, Universitätsstraße 22, 50923 Köln, Germany

18. J.L. Ferrer, Laboratori de Simulació i Investigació Operativa, Departament d'Estadística i Investigació, Operativa, Universitat Politècnica de Catalunya, Pau Gargallo 5, 08028 Barcelona, Spain, lios@eio.upc.es

19. K. Froese, debis Systemhaus GEI, Ulm, Germany, k.froese@debis.com

20. D. García, TSS-Transport Simulation Systems, Tarragona 110-114, 08015 Barcelona, Spain, info@tss-bcn.com

21. C. Gawron, Zentrum für Paralleles Rechnen (ZPR), Weyertal 80, 50931 Köln, Germany, gawron@zpr.uni-koeln.de

22. K. Grobel, Lehrstuhl Informatik im Maschinenbau (IMA/HDZ), RWTH Aachen, Dennewartstr. 27, 52068 Aachen, Germany

23. S. Grosso, Transport Operations Research Group, Dept. of Civil Engineering, University of Newcastle, Claremont Tower, NEI 7RU, Newcastle upon Tyne, UK, Sergio.Grosso@ncl.ac.uk

24. K. Henning, Lehrstuhl Informatik im Maschinenbau (IMA/HDZ), RWTH Aachen, Dennewartstr. 27, 52068 Aachen, Germany, henning@hdz-ima-rwth-aachen.de

25. F. Hofman, Ministry of Transport, Public Works and Water Management, P.O. Box 1031, 3000BA Rotterdam, The Netherlands, f.hofman@avv.rws.minvenw.nl

26. F. Huber, Bergische Universität-Gesamthochschule Wuppertal, Fachbereich Bauingenieurwesen, Lehrgebiet Umweltverträgliche Infrastrukturplanung, Stadtbauwesen (LUIS), Pauluskirchstrasse 7, 42285 Wuppertal, Germany, fh@wbluis.bau.uni-wuppertal.de

27. C.H. Joh, Eindhoven University of Technology, Urban Planning Group, P.O. Box 513, Mail Station 20, 5600 MB Eindhoven, The Netherlands, eirass@bwk.tue.nl

28. S. Kaufmann, Bergische Universität-Gesamthochschule Wuppertal, Fachbereich Bauingenieurwesen, Lehrgebiet Umweltverträgliche Infrastrukturplanung, Stadtbauwesen (LUIS), Pauluskirchstrasse 7, 42285 Wuppertal, Germany

29. C. Kessler, Institut für Geophysik und Meteorologie, Projekt EURAD, Universität zu Köln, Aachener Str. 201-209, 50931 Köln, Germany, ck@eurad.uni-koeln.de

30. M.J. Kerschgens, Institut für Geophysik und Meteorologie, Bereich Meteorologie, Universität zu Köln, Kerpener Str. 13, 50937 Köln, Germany, mk@meteo.uni-koeln.de

31. W. Knospe, Physik von Transport und Verkehr, Gerhard-Mercator-Universität Duisburg, Lotharstr. 1, 47048 Duisburg, Germany, knospe@traffic.uni-duisburg.de

32. S. Krauß, Deutsches Zentrum für Luft- und Raumfahrt (DLR), Linder Höhe, 51147 Köln, Germany, stefan.krauss@dlr.de

33. W. Krautter, Robert Bosch GmbH, Corporate Research and Development, Dept. FV/FLI - Information and Systems Technology, P.O. Box 10 60 50, 70049 Stuttgart, Germany, ARTIST@fli.sh.bosch.de

34. J. Ludmann, Forschungsgesellschaft Kraftfahrwesen Aachen, Steinbachstr. 10, 52074 Aachen, Germany, ludmann@fka.de

35. D. Manstetten, Robert Bosch GmbH, Corporate Research and Development, Dept. FV/FLI - Information and Systems Technology, P.O. Box 10 60 50, 70049 Stuttgart, Germany, ARTIST@fli.sh.bosch.de

36. M. Memmesheimer, Institut für Geophysik und Meteorologie, Universität zu Köln, Aachener Str. 201-209, 50931 Köln, Germany

37. K. Nagel, Los Alamos National Laboratory, Mail Stop M997, Los Alamos NM 87545, USA, kai@lanl.gov

38. L. Neubert, Physik von Transport und Verkehr, Gerhard-Mercator-Universität Duisburg, Lotharstr. 1, 47048 Duisburg, Germany, neubert@traffic.uni-duisburg.de

39. D. Neunzig, Institut für Kraftfahrwesen Aachen, Steinbachstr. 10, 52074 Aachen, Germany, neunzig@ika.d.rwth-aachen de

40. G. Rindsfüser, Institut für Stadtbauwesen, RWTH Aachen, Mies-van-der-Rohe-Straße 1, 52074 Aachen, Germany, rindsfueser@isb.rwth-aachen.de

41. H. Saß, Lehrstuhl Informatik im Maschinenbau (IMA/HDZ), RWTH Aachen, Dennewartstr. 27, 52068 Aachen, Germany

42. R. Schrader, Zentrum für Paralleles Rechnen (ZPR), Weyertal 80, 50931 Köln, Germany, schrader@zpr.uni-koeln.de

43. M. Schreckenberg, Physik von Transport und Verkehr, Gerhard-Mercator-Universität Duisburg, Lotharstr. 1, 47048 Duisburg, Germany, schreck@traffic.uni-duisburg.de

44. T. Schwab, Robert Bosch GmbH, Corporate Research and Development, Dept. FV/FLI - Information and Systems Technology, P.O. Box 10 60 50, 70049 Stuttgart, Germany, ARTIST@fli.sh.bosch.de

45. H.J.P. Timmermans, Eindhoven University of Technology, Urban Planning Group, P.O. Box 513, Mail Station 20, 5600 MB Eindhoven, The Netherlands, eirass@bwk.tue.nl

46. P. Wagner, Deutsches Zentrum für Luft- und Raumfahrt (DLR), Linder Höhe, 51147 Köln, Germany, peter.wagner@dlr.de

47. J. Wahle, Physik von Transport und Verkehr, Gerhard-Mercator-Universität Duisburg, Lotharstr. 1, 47048 Duisburg, Germany, wahle@traffic.uni-duisburg.de

48. H. Wallentowitz, Institut für Kraftfahrwesen Aachen, Steinbachstr. 10, 52074 Aachen, Germany, wallentowitz@ika.rwth-aachen.de

49. N. Wu, Lehrstuhl für Verkehrswesen, Ruhr-Universität Bochum, 44780 Bochum, Germany, Ning.Wu@ruhr-uni-bochum.de

I. Economic Factors of Traffic Generation

Multicriteria Demand Reaction Analysis in Passenger Transport

H. Baum and K. Esser

Institut für Verkehrswissenschaft, Universität zu Köln, Universitätsstraße 22, 50923 Köln, Germany

Current findings of transport policy about its impacts and the reactions of transport demand on pricing or regulatory measures are unsatisfactory. One criticism of previous demand reaction analyses comprises the relatively uniform modeling. However, transport participants react on changes of traffic conditions quite differently. This study aims at going beyond these uniform approaches and at working out the variety of transport demand reactions.

1 Introduction

The success or the failure of transport policy measures to improve traffic conditions depends on the reactions of road users, i.e. changing of transport behaviour, acceptance and application of policy measures by people and economy. Information about the impacts resulting from measures of transport policy are of essential significance for political decisions because new measures and instruments are planned and discussed both in the national and in the European transport policy. However, practical experiences for estimating their impacts can not be provided. This is true, for instance, concerning fiscal regulatory measures (e.g. road pricing), new technologies in transport (e.g. 3-liter-car, natural gas vehicles), telematic applications (e.g. new information and communication technologies) and measures to rationalize transport performance.

Simulating transport performance and its environmental impacts that covers the interactions between transport participants´ behaviour and transport policy measures, needs for a better insight to individual demand reactions. Hence, the restricted transferability of one-dimensional reaction analyses should be avoided. This is done - inter alia - by new methods of reaction analysis that helps working out a broader spectrum of reactions and impacts. This multidimensional reaction analysis of transport demand provides the needed input to simulate traffic. In so far, it is necessary to focus on the economic analysis of transport demand.

2 Methodical Approach

2.1 Restrictions of Previous Demand Reaction Analyses

Transport participants´ reactions have to be respected within demand reaction analysis of policy measures to relieve traffic and environment. These reactions perform the essential link between the implementation of a measure and its impacts on traffic. Previous research provides no sufficient data bases and reaction indicators. Therefore, reliable information about the demand reaction following pricing measures are limited.

The reasons for this result from various topics:

- In most cases reaction analysis so far proceed only one-dimensional. However, estimating the impacts of transport policy measures should not be based on uniform effects. Moreover, a bundle of reactions should be regarded. Transport participants are - to certain degrees - free when reacting on changes of transport´s frame conditions.
- The other way round, specific impacts are explained by single policy measures, while various measures in cooperation really cause a certain reaction.
- Additionaly, traditional reaction analysis (e.g. econometrical regression analysis) cannot determine the impacts of measures which have not been introduced yet.

Analyses of transport demand so far is output-oriented because only a result is derived. It is not clarified, by which this result has been initiated (Fig. 1).

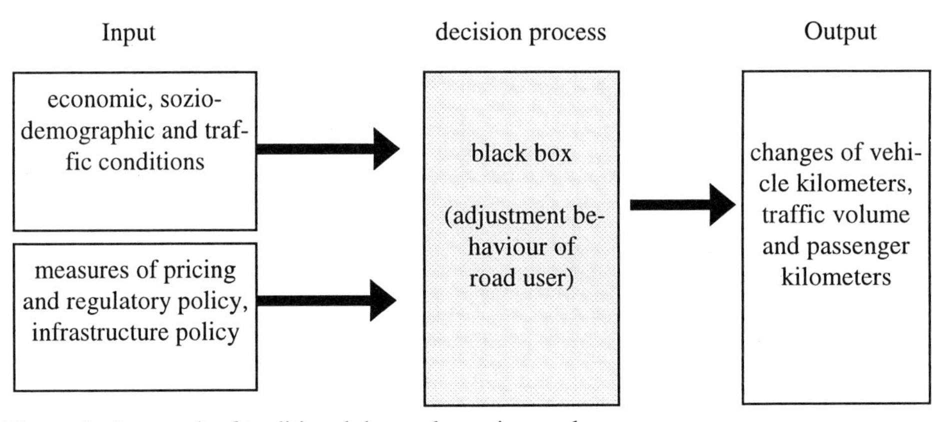

Figure 1: Approach of traditional demand reaction analyses.

The decision process (black box) of a road user is not exposed. Because of not revealing the deciding individual adjustments, the impacts on traffic are often over- or underestimated.

2.2 Further Development of Methods of Reaction Analysis

Our investigation aims at coming off uniform reaction analyses and at demonstrating the multidimensional reaction patterns of road users. For this purpose, the decision process (black box) is exposed.

The following adjustment measures are worked out:

- Changes of traffic generation and modal choice.
- Measures of organization and rationalization.
- Acceptance and application of telematic systems.
- Restructuring of consumption expenditures.
- Downsizing of vehicle fleet.
- Acceptance and market penetration of new vehicle technologies.

Special focus is laid on the integration of accepting environmental friendly and traffic-saving measures into the transport demand analysis (e.g. acceptance of new vehicle technologies, acceptance of telematic systems). It is obvious that transport policy´s options do not fit to transport participants capabilities and doings. Analyzing acceptance will be the key to a successful transport policy in the future.

The analysis of acceptance and market penetration of measures to relief traffic and environment provides new findings about transport participants´ reaction patterns. Furthermore, these insights to acceptance contribute to decover the potential addressing road users´ mobility decisions. This study informs to which extent increasing acceptance towards certain strategies may exploit traffic and environment improvements.

Different methods of reaction analysis in passenger transport are used in this study.

- For certain measures, an enlarged range of reaction patterns is defined and analized empirically. Especially intermediate results - leading to a definite demand - are worked out. Thus, the decision process of transport participants should become more transparent. For estimating this, conventional methods of inductive statistics are used (e.g. regression analyses).
- Methods of the business economies (break even-analysis) are applied to transport demand in addition to conventional approaches. These cost-based decision models determine the reactions of transport demand, e.g. on pricing or regulatory measures, by means of cost comparisons. Quality differences between the alternatives are transfered into cost equivalents. Hence, an alternative to conventional reaction analyses is offered.
- New and additional information about transport behaviour of road users are achieved by developing a general model for transport demand and its application on selected alternative demand reactions. Likewise, the findings of conventional methods can be tested.

An important shortcoming of previous demand reaction analyses - i.e. considering mono-causal impacts - is avoided by the multicriteria approach. Its results represent the input data to the simulation model, covering road users´ behaviour and reactions on single traffic performances or measures. Selected results in passenger transport are presented in the following chapter.

3 Empirical Results

The economic analysis of transport demand is realized by selected strategies and measures of transport policy. Their impacts are estimated and goal achievement is tested.

3.1 Changes of Traffic Generation and Modal Choice

Transport participants react on pricing measures by reducing their mobility. The dimension of reduced transport demand is revealed by elasticities. Tab. 1 shows elasticities (empirically estimated) of road transport and public transport based upon surveys.

	Dependent variable	
independent variable	vehicle kilometers in passenger transport	Passenger kilometers in public transport
price and income elasticities		
• fuel price	-0.25	0.40
- household, low income	-0.36	
- household, medium income	-0.28	
- household, high income	-0.20	
• real income	1.40	1.12
• price of public transport		-0.25

Table 1: Price and income elasticities (empirically estimated).

In general increasing fuel prices only lead to relatively low reductions of the vehicle kilometers. The low price elasticity states the high preference of road users towards car traffic. The intensity of reaction depends, among others, on passengers´ level of income. The higher the income, the lower the impacts of fuel price. Households with a low income react stronger (-0.36) than households with a medium (-0.28) or with a high income (-0.20).

The initiated shifts to other modes by means of increasing fuel prices are comparatively small as well. The changes of the modal split are estimated by means of cross price elasticities. The cross price elasticity amounts to 0.40. In contrast road pricing has got stronger impacts on vehicle kilometers than fuel prices. The incidence of user charges is stronger for their monetary burden is directly and more differentiated realized by road users.

3.2 Effects Resulting from Measures of Infrastructure Policy

Mobility takes place on existing transport infrastructure. Transport behaviour is influenced, among others, by transport infrastructure´s quality. Construction and extension of road infrastructure improves the traffic conditions in road transport.

There is a controversy about infrastructure improvements and capacity increases because new transport demand is generated. Thus the success of such strategies is limited that could contribute to decreasing transport volume. Different opinions exist about dimension, level, and relevance of induced traffic. They result from differing definitions of induced traffic.

The relationship between the expenditure for road infrastructure and vehicle kilometers is estimated for the whole road network and for single road categories by means of quality elasticities. They describe the impacts of improving road infrastructure on vehicle kilometers in passenger transport. The corresponding results are given in Tab. 2.

	Dependent variable	
independent variable	vehicle kilometers in passenger transport	vehicle kilometers on federal trunk roads
quality elasticities		
• gross capital assets of road infrastructure	0.50	
• length of road network	0.28	
• gross capital assets of federal trunk roads		0.36

Table 2: Quality elasticities (empirically estimated).

Following this, an extension of road capacities results in a greater volume of traffic. The effects on vehicle kilometers in passenger transport are explained by the extension of road infrastructure. The additional effect arises from improvements of the existing road infrastructure´s quality (e.g. widening lanes or adding lanes on motorways).

3.3 Acceptance and Application of Telematic Systems

Using navigation and route-guidance systems, transport participants can individually benefit from time cost and vehicle operating cost savings (e.g. by-passing congestions). Such systems broaden road user´s reacting patterns (following price increases) and enable them to realize their mobility needs more efficiently.

Both acceptance and application of telematic systems are influenced by the individual benefits road users are able to realize. By means of a break even-analysis, the mileage per year is determined that represents the margin for a profitable investment into a dynamic route-guidance system. The break even-analysis is represented in Fig. 2 each for a cost saving of 5% and 8% (time costs and vehicle operating costs).

Cost decrease or increase respectively of integrated telematic systems DM/year

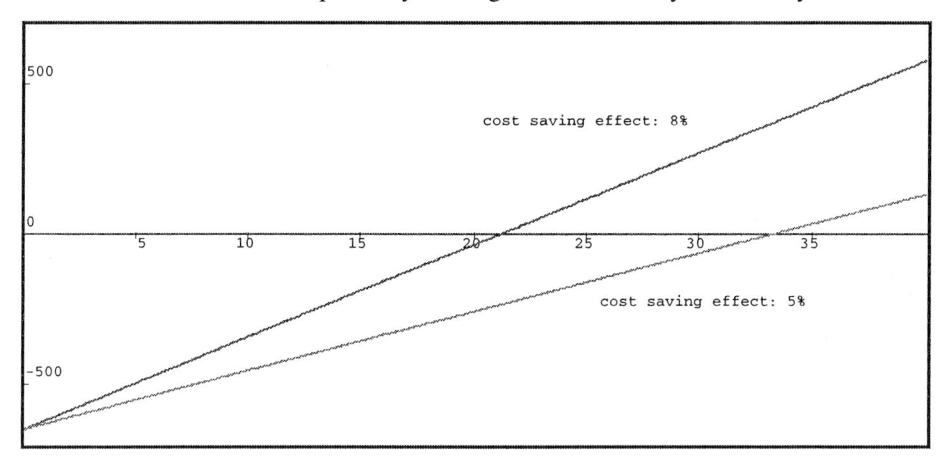

vehicle kilometers (in 1,000 /year)

Figure 2: Break even-analysis of dynamic route-guidance systems.

The application of a dynamic route-guidance system is profitable beyond 33,275 vehicle kilometers (5%) resp. 21,198 vehicle kilometers (8%) per year. Increasing fuel prices show positive impacts on the acceptance and the benefits of telematic systems. The application of such dynamic telematic systems will be profitable beyond 28,247 vehicle kilometers (savings: 5%) per year by increasing of fuel price by 50%. That means, a decline of break even by 5,027 vehicle kilometers.

The break even-points have been calculated here partially explain the low ratio of application up to now. On the one hand, telematic systems are (still) too expensive. On the other hand, the individual benefits for each vehicle user are too little with the corresponding investment costs.

3.4 Restructuring of Consumption Expenditures

Road users are able to compensate the monetary burden of pricing policy (e.g increasing of fuel price, road pricing) by restructuring their consumption expenditures. If the share of transport expenditures increases, the share of other consumption expenditures decreases.

The potential to restructure the transport expenditures (with reference to all households) is described exemplarily from 1990 to 1994 (Tab. 3).

year	average increase of (real) expenditure for individual traffic	thereof realized through:		
		increase of real income	restructuring of consumption expenditures	restructuring of transport expenditures
	2.85%	2.17%	0.48%	0.20%
	(billion DM)	(billion DM)	(billion DM)	(billion DM)
1990	5.184	3.947	0.883	0.354
1991	5.683	4.327	0.978	0.378
1992	5.568	4.240	0.946	0.382
1993	5.050	3.845	0.831	0.374
1994	4.992	3.801	0.821	0.370

Table 3: Sources of financing the growth of real expenditures for individual transport (1990 to 1994).

Between 1990 and 1994 the real expenditures for individual traffic have been increased on average by 2.85% and the real income on average by 2.17%. That means, in 1993 0.68% (1.2 billion DM) of expenses belonging to individual transport have been covered by restructured consumption behaviour and 2.17% (3.8 billion DM) have been paid by increases of real income within this period.

It can be concluded that in 1993 an additional burden (e.g. as a result of road pricing) of 1.2 billion DM could be compensated by restructuring of consumption expenditures. With reference to vehicle kilometers on motorways (151.5 billion vehicle kilometers in 1993) it will be thus possible to bear a motorway charge of approximately 1 Pf (0.8 Pf) per kilometer by changing consumption behaviour. Furthermore, road users are able to finance motorway charges up to the amount of 2.5 Pf per kilometer by the increase of real income.

It turns out that the impacts of motorway charges are difficult to estimate. An analysis of the effects of pricing policy on transport demand (e.g. concerning changes of vehicle kilometers) can only be realized considering the saving potential which is given by restructured expenditures.

3.5 Downsizing of Vehicle Fleet

Downsizing of vehicles is a possibility to partly compensate the effects of pricing policy in individual traffic. Buying a smaller and less motorized vehicle decreases vehicle price, fuel consumption, and other costs (e.g. vehicle-tax, insurance premium) without additional technologies. This reaction pattern has mostly been neglected in previous demand analysis. Nevertheless, the information about structure as well as changes of the vehicle fleet is necessary to simulate traffic flow and its environmental impacts.

The impacts of an increased fuel price on the average cubic capacity (cc) of new vehicles licenses are estimated exemplarily. The change of average cubic capacity of new vehicles by increasing fuel prices (e.g. 50% and 0.20 DM per liter) is shown in Fig. 3.

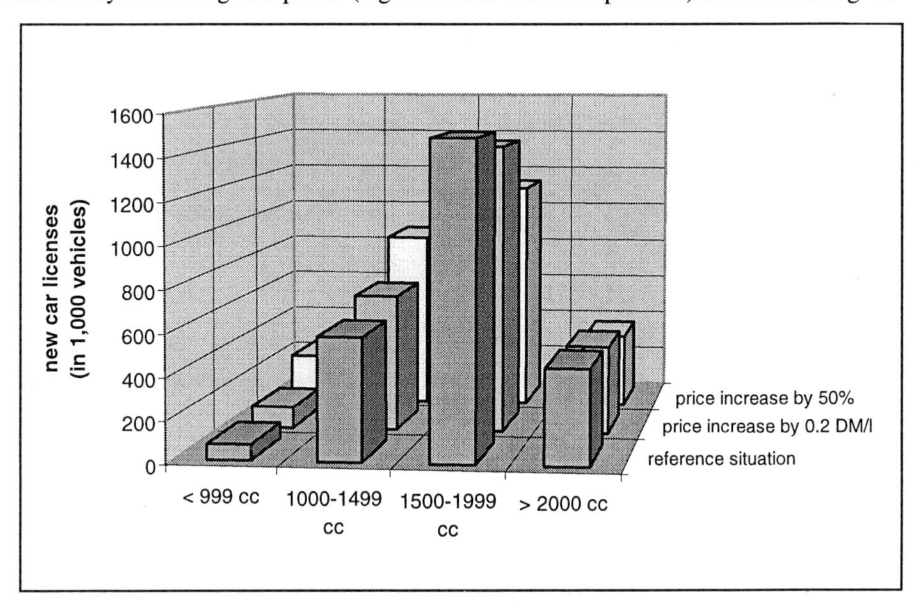

Figure 3: Impact of increasing fuel prices on cubic capacity of new vehicles.

An increase of fuel prices by 50% leads to the following effects:

- The share of new vehicles up to 999 cc increases by factor 3.
- The share of new vehicles between 1,000 cc and 1,499 cc increases by factor 1.5.
- The average cubic capacity of a new vehicle changes from 1,732 cc to 1,559 cc.

It has to be stated that the share of smaller and less motorized vehicles increases remarkably by rising the fuel price. Road users´ willingness to adjust their behaviour like that will decrease in the following years, for instance resulting from real fuel price decreases, from becoming used to a higher level of fuel prices or from increasing incomes.

3.6 Acceptance and Application of New Vehicle Technologies

Technical innovations are of essential significance to preserve mobility and to relieve environmental sufficiently. The industry is more and more involved in developing new vehicle technologies. Using energy-saving technologies (especially 3-liter-car) is qualified to relieve environment and to reduce carbon dioxide emissions.

The success of a shifting of the existing vehicle fleet to low-consumption and low-emission vehicles depends decidingly on road users´ acceptance. Therefore, it has to be worked out which kind of traffic can be done by 3-liter-cars, and how many vehicle kilometers per year can be shifted to 3-liter-cars. Current findings about the potential usage of new technology vehicles serve as the basis for estimating traffic and environmental relief.

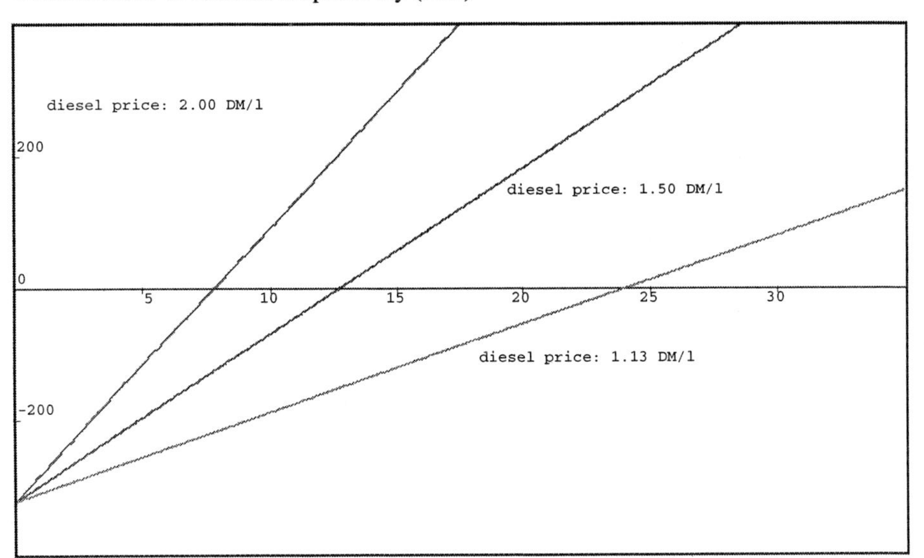

Figure 4: Break even-analysis of 3-liter-car (revealing a price difference of 2,000 DM).

The break even-point of buying 3-liter-cars is calculated by a comparison of costs. This comparison between a conventional car and a 3-liter-car includes, among others, the vehicle operating costs, the capitalized price, and costs resulting from differing qualities as well (Fig. 4).

The use of a 3-liter-car is just advantageous upwards of 23,945 vehicle kilometers per year (considering the current price of diesel fuel). According to these preconditions, the 3-liter-car is no complete full alternative to a comparable small car.

An improvement of acceptance could be reached by pricing policy measures. Within 10 years, a strong increase of the diesel price (by about one third) through rising the fuel tax could result to substituting about 21% of vehicle stock and 23% of vehicle kilometers by 3-liter-cars.

4 Conclusions

A further development of methods and the application of cost-based decision models enable a better insight into road users´ reaction patterns. Traditional demand reaction analyses only considers mono-causal impacts. By contrast, the method presented here is a multicriteria reaction analysis and is therefore able to avoid partially the imperfection of previous approaches.

This study points out possibilities and limits of pricing measures in passenger transport. It provides information about the possibilities to influence and to direct transport demand as well as about the "correct" dosage of pricing measures. The empirical as well as practical frame given through this study can serve as assessment tool for political decision makers.

II. Traffic Generation

Dynamic Estimation of Transport Demand: Solutions – Requirements – Problems

K.J. Beckmann and G. Rindsfüser

Institut für Stadtbauwesen, RWTH Aachen, Mies-van-der-Rohe-Straße 1, 52074 Aachen, Germany

The following contribution presents requirements, difficulties and first attempts at modelling a temporal short interval estimation of transport demand. For the simulation of motorway traffic (for an area within the state of North-Rhine-Westfalia, NRW) methods of temporal disaggregation of existing trip-matrices have been worked out within the framework of the Northrhine-Westfalian research cooperation for traffic simulation and environmental impacts "NRW-FVU" (Nordrhein-Westfälischer Forschungsverbund Verkehrssimulation und Umweltwirkungen). To simulate urban traffic (example: Wuppertal) methods of estimating temporal short interval trip-matrices were conceived and tested. The matrices were supplied for microsimulation with Cellular Automat (CA) and for the dynamic route choice and traffic assignmemt (DRUM, Dynamische Routensuche und Umlegung). The comparison of both methods, based on the estimated link loads (ADT and hourly loads), supplies deviations ranging within the mean variation of counted values. It therefore can be inferred that these methods, which are different with regard to computing intensity and data requirements, should be used depending on the tasks and the intended precision of the results. The specific pros and cons are important operational criteria. It also becomes obvious that in future methodical advancements should be examined on the basis of activity(-chain)-based approaches.

1 Introduction

Among other things, the aims of the interdisciplinary research cooperation are determined by the efforts of the government of the state of North-Rhine-Westfalia, to preserve mobility, to improve the efficiency of transport systems and to reduce noticeably negative traffic impacts.

In the planning practice of municipalities, regions and states, these are common targets for the organisation of transport systems. Measures of traffic avoidance, modal (traffic) diversion, system managment as well as short term on-line traffic control (Fig. 1) could equally make a contribution to these aims.

Only with such integrated measures, it can be guaranteed that traffic does not become the limiting factor of social and economic development. The various temporal horizons of measures and different measure levels necessarily have to be taken into account. These measure levels are:

- Strategic Planning:

The measures of Strategic Planning tie down (Transportation politics, regional- and infrastructure politics) a considerable amount of social resources. The approaches only have effects over the medium and long term. In most cases a short-term achievement is not feasible. Furthermore, the approaches of strategic planning are indispensable to the protection and improvement of the effectivenss and efficiency of system control as well as online control. The modelling of transport demand – i. e. the estimation of traffic volume (trip generation), the spatial trip distribution (intraspatial transport demand) and the modal split – thus is of great importance. The resulting trip matrices are the input data for route choice and traffic assignment to respective traffic networks.

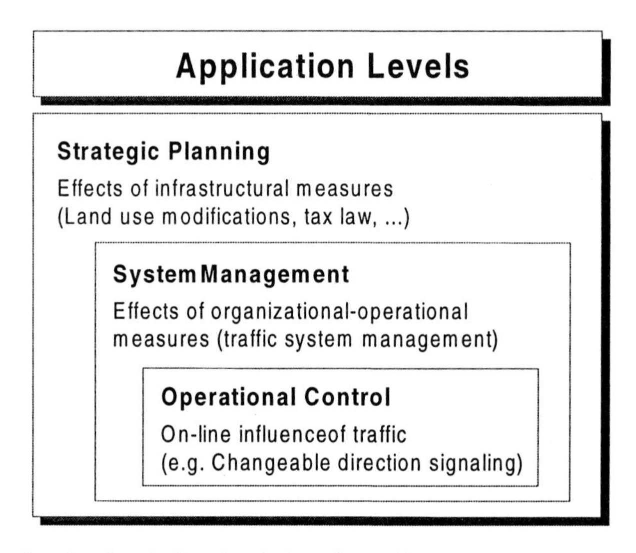

Figure 1: Application levels for simulations in traffic planning.

- System Control/System Management:

Within the framework of system control (transport system management), medium- and long-term control strategies are developed and tested. Among others, these strategies are the fixing of service qualities (such as fixed time) and fares within the public transport, of part-spatial admissions of means of transport (modes), of fundamental speed regulation, or of parking management. The measures of system control make possible an efficient and ecologically harmless use of the road infrastructure – and therefore of social ressources. In this context, the modelling of transport demand offers the basis for the estimation of the effects of these measures as far as the satisfaction of transport demand (preservation of mobility), the efficient use of existing infrastructure and the reduction of undesirable effects of traffic are concerned.

- On-Line Control:

Within the framework of on-line control, short-term strategies for complete utilization of capacities and, therefore, for an increase of the efficiency of the road traffic system are designed, tested, and put into force. An efficient on-line control may be based on a transport demand model, which determines traffic volume, trip distribution, modal-split, and traffic assignment in short temporal intervals (dynamic modelling). An on-line control on the basis of short-term traffic condition studies can also be carried out. To insure stable and foreseeing control strategies, on-line informations should be linked with „dynamified“ transport demand calculations. Those measures should result in learning-based models for the description of sudden changes of traffic flows in parts of the network. The determination of base loads in short temporal intervals with the help of transport demand modelling is therefore essential for such models, especially in cases of long-term estimation of traffic flows. Hence, on-line control serves the short-term and part-spatial increase of the street network´s efficiency and the reduction of part-spatial peaks of environmental impact caused by traffic. However, this does not ensure a basic guarantee of mobility through efficient use of the infrastructure and through the fundamental improvement (reduction) of environmental impacts.

In this context, transportation demand modelling and traffic simulation serve to prepare decisions of medium- or long-term effectiveness by means of prognosis of model inputs (e.g. population structure and road behaviour data) as well as by means of simulation of the impact of various measures on traffic (Fig.2).

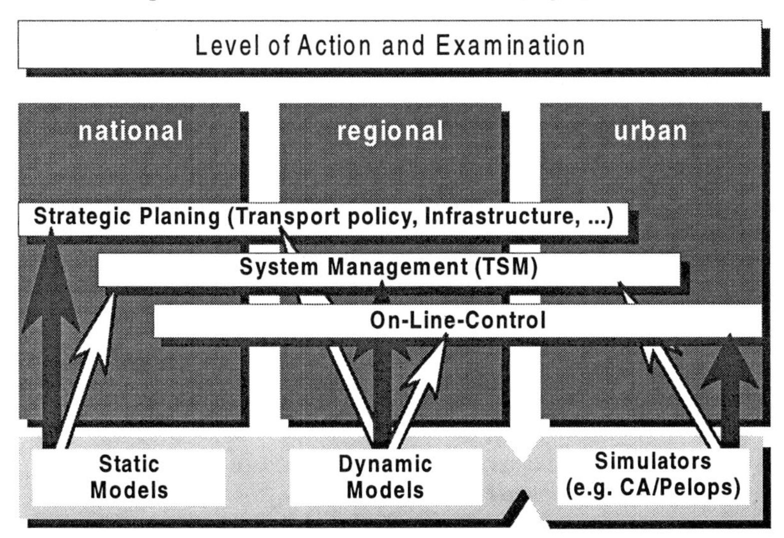

Figure 2: Levels of measures and assignment of models.

With reference to the behaviour of individual travellers or those demanding for traffic services (shipper, carriers), measures of strategic planning have, among others, influence on:

- the individual behaviour concerning locational choices (residential locations, workplace locations, industrial locations),
- fundamental structures of individual activity patterns and time-space-behaviour as well as on economic exchanges,
- patterns of travel behaviour (traffic volume, modal preferences),
- the availability of individual modes of transport.

Measures of the **system control**, however, affect the individual transportation behaviour on the meso-level, i.e. routine decisions which take place before departure/the start of the journey and which are relatively stable in time, such as:

- the choice of optional activities,
- the locational choice for regular activities,
- modal choice,
- the choice of courses of activity and chances of travel behaviour.

Measures of **on-line control**, however, work exclusively (or at least predominantly) traffic system immanent, i.e. they have virtually no influence on the time-space-behaviour regarding the choice of activity, the locational choice of activities, and the time of departure. Influence on the choice can only be effective if the information about traffic conditions is made available sufficiently in advance of the time of departure (pre-trip). In principle, only the short-term choice of routes in a traffic network is influenced, such as individual ways of reaching a destination, and at best there is a small influence on the choice of and linking of modes of transport (e.g. Park and Ride). On-line control helps to reduce the probability of incidences of overloading by way of route and, if applicable, speed recommendations for a given network. If effective and long-lasting measures are to be taken, to fundamentally preserve mobility, to improve use of the whole traffic system (all road-users included), and to improve the effects caused by traffic, the methodology of modelling traffic demands on all levels of consideration must be broadened and improved. Improvements are especially required in

- introducing estimations of the change of road users behaviour, with changes made to basic traffic conditions (e.g. structure of costs) into traffic demand modelling,
- the amount of time taken to carry out estimations of traffic demand,
- extending basic principles to include the consideration of freight traffic.

Within the area of speciality of the Institute of Urban and Transport Planning (ISB, Institut für Stadtbauwesen) in NRW-FVU, the conception and testing of methods for estimating traffic demands were the main areas. At the same time, the project was split into two areas of application.

1. Simulation of traffic in a subnetwork of the North-Rhine-Westfalian motorway network and

2. Simulation of city traffic with the illustrative town of Wuppertal.

The aim of the ISB was to prepare vehicle-specific trip matrices for passenger transport over a long period of time, to be used as the basis or input for dynamic route choice and traffic assignment (DRUM [1, 2]) and a microscopic simulation of traffic flow, with the help of a Cellular Automat (CA, see [3, 4, 5]).

2 Division of an Extensive Trip Matrix for the Simulation of Motorway Traffic with Respect to Time

Targets/Requirements

The aim of this project was to produce a temporal short-interval representation of traffic occurring on motorways in a specific area of North-Rhine-Westfalia. The daily traffic on the motorways was to be simulated with the help of the Cellular Automat. Connections of sources and destinations were used in short time intervals as input data. As the CA works with individual vehicles, in general it should be possible to allocate a starting point in time to each vehicle.

The ADT-trip matrices normally used in the area considered or the usual matrices for the four-hour peak group (which may occur over the peak hour) are therefore rather unsuitable as input data, or at least as the basis of the input for the CA. A further requirement is the ability of the models to illustrate measures taken concerning traffic. This requires a methodology for the preparation of trip matrices, to be worked out over a period of one day in at least one-hour intervals (if possible in half-hour intervals).

In Addition to that, it should be clarified during the preparation to what extent this methodology is suitable for forcasting traffic (on a long-term basis, as well as medium- and short-term ones).

Methodology

Usually, sequential traffic models (e.g. Four-level algorithms) are used to calculate trip matrices on a regional and urban level, which determine traffic demands from information about structures (population distribution, work-places etc.) and behaviour (e.g. specific revenue figures or travel rates), then examine the traffic distribution and calculate modal proportions. Route choice and traffic assignment are applied next. (see e.g. [6, 7]).

The methodology and models available are validated and produce realistic results for the area of their application. Generally, such an approach can be transferred to regional level. The requirements concerning the detail and range of input data, however, are significant. Therefore, on a regional level, aggregated approaches are preferred. These cannot be modified for set requirements, as most are not behaviour-orientated.

As for these reasons and because of the qualifications and basic conditions within the research co-operation, a methodology was worked out for the division of an existing ADT-matrix (from studies forming part of the national traffic plan of North-Rhine/Westphalia in 1990), to prove general workability of comprehensive model run-throughs. A theoretical matrix was decided against, as apart from interface checks on the Cellular Automat, above all plausibility checks were planned.

The following work steps were carried out in detail, to divide the matrix:

- The matrix was checked for completeness and the format of the data was adapted to the Cellular Automat for the interface demand estimations (ISB).
- The matrix was updated. To achieve this, counted values collected from a field test carried out by the research co-operation were compared to the values of the automatically continuous counts in the motorway network of the investigation area, and corresponding development factors with regard to time were compiled as an adjustment.
- The links between source and destination given in the matrix were reduced to a trip matrix from motorway exit to motorway exit using route choice and traffic assignment as well as route backtracing. The group of trips was split up into the subordinate networks of federal, regional, county, local and other roads.
- The matrix was divided further in time intervalls of 1 hour. In this investigation, this was done by using a time graph for different trip purposes, whereby the proportion of the matrix for each trip purpose was estimated, in order to be able to illustrate the specific characteristics of the trip purposes with regard to time. At the same time, the ADT values were sorted into either four-hour or peak hour values for a plausibility check, using factors taken from research reports [8]. Furthermore, factors were derived from numerical data obtained in the field test in five-minute intervals, which allowed the theoretical formation of matrices in five-minute steps. During the division, however, the problem arose that 'direction surplus', i.e. the time-dependent differing parts of the ADT in various directions, must be represented. The temporal short interval trip matrices required for a later point in time cannot be formed as a source-destination matrix from one traffic cell to another using this procedure, as only inadequate conclusions can be drawn about actual trip relations from roadside counts available (see also FGSV [9]), and such methods especially can rarely be used for forecasting.

Using this procedure, trip matrices in one-hour intervals were made into ones over a whole day, which were used as input data in load estimations with the CA and DRUM. To compare the daily loading of the routes, the daily traffic was estimated with the help of a static route choice and traffic assignment, using added-on matrices.

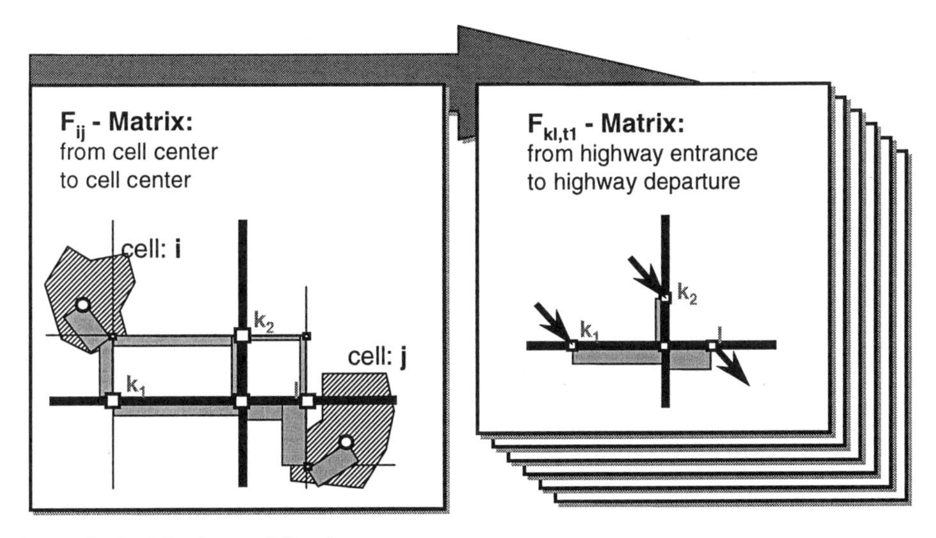

Figure 3: F_{ij}-Matrix modification.

Results

The results presented here are to be regarded as provisional since comparisons have not yet completely been made. In addition to that, the methodologies of the procedures are being rigorously tested at present.

regarded difference	Absolut [difference related to first addend]				Proportional [% related to first addend]			
[1]-[2]	Min [1]-[2]	Max [1]-[2]	Average abs ([1]-[2])	Deviation abs ([1]-[2])	Min [1]-[2]	Max [1]-[2]	Average abs ([1]-[2])	Deviation abs ([1]-[2])
Fe-BA	**2.635**	**13.462**	**5.665**	**3.603**	**10.67**	**93.96**	**24.22**	**24.57**
DRUM-BA	**-27.288**	**23.318**	**9.868**	**3.406**	**-88.63**	**91.20**	**29.41**	**23.08**
CA-BA	**-42.588**	**45.857**	**11.929**	**9.583**	**-84.65**	**170.36**	**35.89**	**31.10**
DRUM-Fe	**-18.870**	**10.903**	**12.273**	**3.804**	**-72.13**	**29.60**	**44.37**	**18.38**
CA-Fe	**-24.713**	**8.246**	**11.442**	**6.412**	**-71.07**	**22.39**	**41.34**	**23.51**
DRUM-CA	**-46.118**	**48.923**	**10.034**	**10.655**	**-89.29**	**331.69**	**30.90**	**40.56**
Fe = Count Values during a special time period / BA = Count values (permanent counting points BASt) DRUM = Dynamic route search and assignment / CA = Cellular-Automaton [n=216 net elements with permanent counting points (BASt), (10 Fe counting points)								

Table 1: Comparison of loading values (absolut and proportional).

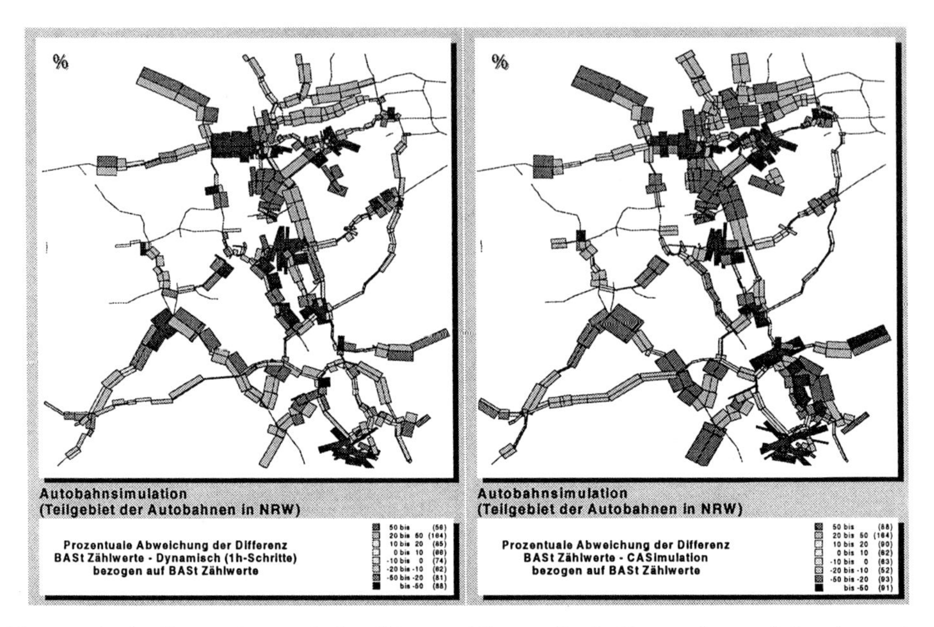

Figure 4: A Comparison of the CA-simulation with the BAST count values

Figure 5: A Comparison of the dynamic route search with the BAST count values

By comparing the loading values (summed up firstly over a day, see Tab. 1) obtained from the calculations using the automatically continuous counted values supplied by the Federal Institution for Roads, BASt [10], it can be seen that for the main part, the models differ slightly from one another. These deviations are usually below 20 per cent. On the other hand, the deviations of the individual methods in relation to the figures are larger. Although it should be taken into account that the BASt counted values represent annual averages, the area of scatter of the figures is considerable and lies in the order of the most frequent deviations.

It should also be recognized that DRUM and CA produce clearly differing loadings in relation to counted values in some areas. Over the duration of the day the procedures generate a higher loading on routes that can be regarded as alternatives to heavily loaded sections. This can be explained from the methodology´s point of view, because now, in place of a single traffic procedure (static method), for each step in time the resistance in the network is determined, the best routes are sought and the trips are divided among the routes. Despite the adoption of different (adapted) capacity functions, alternative routes, which were possibly more lightly loaded, were more frequently allocated in the dynamic procedure. For a more exact analysis see [11].

Fig. 4, 5, 6 and 7 illustrate the comparisons of the traffic flows in the given combinations.

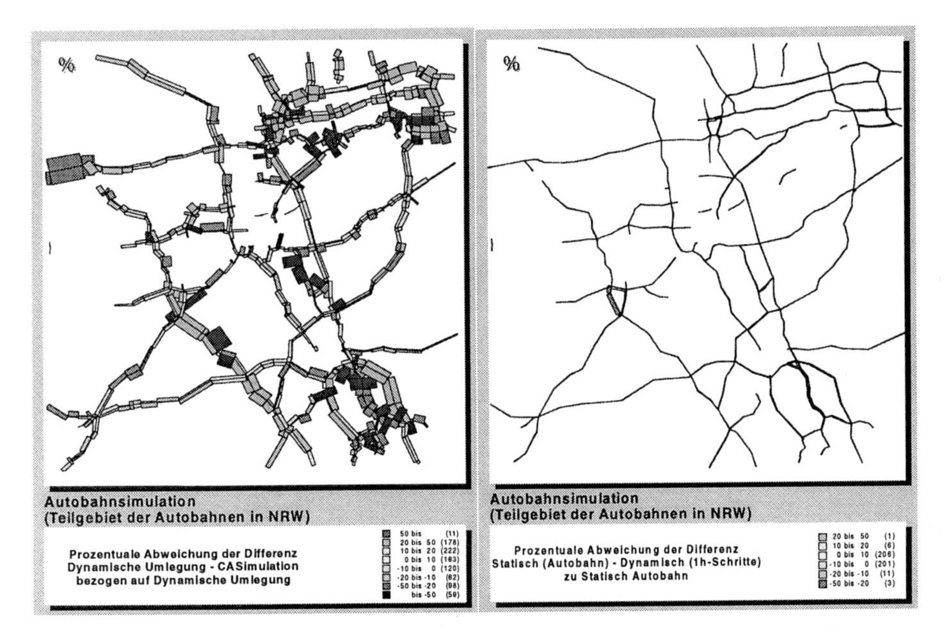

Figure 6: A Comparison of the dynamic traffic assignment with the CA-simulation

Figure 7: A Comparison of the static with the dynamic traffic assignment

In the analysis of the daily pattern over time, the counted and the calculated patterns over time with the help of DRUM and those calculated ones with the help of CA, were overlaid graphically for the route elements of the network next to which lay BAST-continuous census points. The results of the data resulting from the field tests (research co-operative survey June 1995) were illustrated additionally. In Fig. 8 the overlaying patterns over time for chosen transverse profiles are presented. Looking at all present census points it is obvious that the courses during the day are very well recreated by the simulation methods DRUM and CA. Less loaded track sections offer a higher deviation as high loaded ones. All in all we can say that the ranges between counting value and simulation during the micro-simulation with CA is higher than during the dynamic assignment DRUM (Tab. 1).

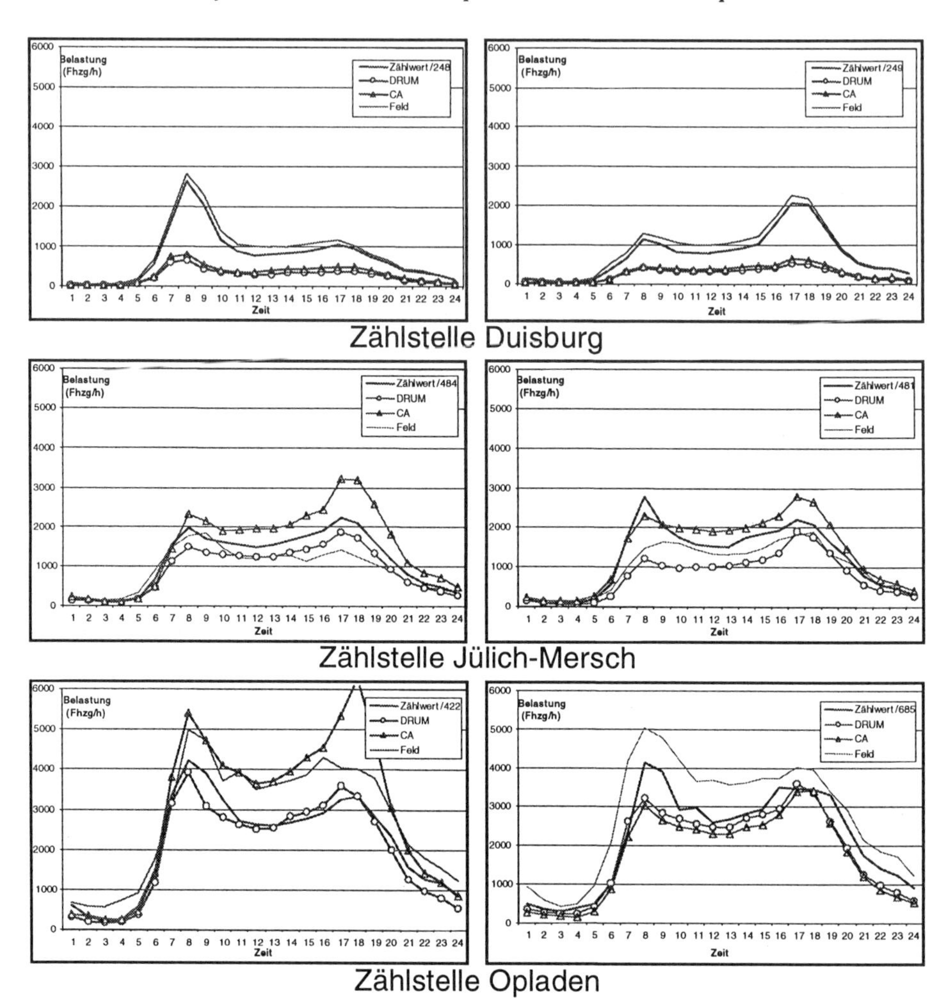

Figure 8: Overlaying patterns over time in 1 hour steps for three chosen transverse profiles.

We have to take into account that BASt-counting values represent an annual average. The deviation of one single day can be read from the pattern time of the field's test counting value. The field test was carried out in June 1996 by the partners of the research co-operative. At this, deviations between the daily values of one week of about 20 % were coming up around the average of the counting values. Therefore, in the chosen method, simulated load's courses can absolutely be classified as satisfying in the conclusion.

These results are based on the runs executed on the first model and initial model settings - no parameter settings were varied at first.

Assessment

The differences in the traffic loading diagrams (comparing the methods over the values from one day) caused by the different procedures give no indication of the quality of the procedures. The deviations in counted values vary with procedure in different ways - there is no recognizable system. Thus, it can be stated that for the illustration of the extensive (motorway traffic on a regional level) action of traffic, both the static and dynamic procedures yield useful results - providing the input data can be prepared for the dynamic procedure. For all applications (e.g. planning problem), for which the consideration of time is not primarily necessary (such as e.g. long term infrastructure management), the cost, judged by the gain in precision of statement, is relatively high and hardly justifiable.

An advantage of the dynamic procedures, however, in particular the CA, is the illustration of traffic action at each point in time. Above all, this advantage becomes meaningful if, for example, the effects of changeable direction signs in the motorway network, or the effects of individual routings are to be investigated.

A validation of the quality of illustration of the CA in the area considered, however, could only be performed to some extent within the limits of the research co-operation (see [11]). Statements are barely possible, particularly due to the lack of relevance of the original matrix, but also due to the differences between the temporal references of the matrix, the BASt counted values and the field test data, as well as the isolated survey during the field test.

The application of dynamic procedures is necessary for system control (medium term - long term), as measures to be taken in this field will intervene directly in the action of traffic. At the same time, knowledge of source-destination connections is particularly necessary here, (although less-so on the level of on-line control) to produce meaningful control strategies and route recommendations, which are of real use to the driver (or to the general public).

The dynamic models probably cannot be employed as a control foundation, exclusively on the basis of the calculated trip matrices, since the input parameters for the demand estimation models are too undifferentiated regarding time. In the future it should be examined to what extent a combination of these models can be used to calculate the base load (existing for different temporal and spacial areas) with the procedure of matrix definition, from real roadside counts (on-line), as a foundation for control strategies (possibly 'learning' models).

3 A Temporal Detailed Calculation of O-D Matrices for an Urban Area

Aims / Requirements

Regional and urban levels are the main operational area for traffic models. The necessity to show the most different measures on all levels concerning their effects, becomes more and more important in megalopolises because of the increasing traffic problems. Resulting from the experience already made with the decomposition of matrices for a wide ranging traffic simulation, Wuppertal was chosen for the development and the verification of the model chain of the NRW-FVU for this special case of application. Concerning the motorway simulation, the primary aim was to carry out a micro-simulation for a wide net and in doing so, to verify the general capability with the help of matrices "close to reality".

The methodology of the calculation of trip generation can be seen as the main focus of all of the work on the urban level, since, here, usually models are used which allow to prognose the road behaviour and hence the traffic flows on the route.

Therefore, the aim can bc defined as the modification of the present methodological starting point with regard to the temporal disaggregation and the allocation of temporal detailed O-D matrixes for micro-simulation (CA) and for a dynamic route choice and assignment (DRUM).

Methodology

At ISB, a disaggregated person-category approach [12] is used for the calculation of trip generation. This model is composed of a sequence of steps:

1. Trip generation

   On the one hand, the amount of home-based trips on the active side are calculated by multiplication of the units of the person-categories with the respective trip rate; on the other hand the home-based trips on the passive side are calculated using the units of the structural size (e.g. employment, school places) and the respective rising value. Afterwards, the Modal Split I is estimated. Here, the respective amount is split up into three partial collectives, namely the bound to the MIV, the bound to the public transport and the facultative collective.

2. Trip distribution

   With the help of a gravity approach, the distribution of the traffic is generated in a specific area concerning the respective aims. With a method of compensation, the calculated matrices are to be modified iteratively over the sum of columns and lines, until the deviation undergoes a fixed limit.

3. Modal Split II (trip division)

   The distribution of the facultative travellers is calculated with the help of a decision-making model (probit and logit-methods are used).

4. Route choice and traffic assignment

   The traffic flow on the route is estimated using a dynamic method for route choice and assignment (DRUM). This technique can also be used as a static one.

With regard to the possibility of realising temporarity detailed calculations, the modification of these models was gradually undertaken during the time of elaboration of the basic data or of the basic size for different models and calculations.

Trip Generation (with Modal Split I)

A huge empirical basis, which can only be established via expendable questioning, is the prerequisite for a temporarity detailed trip generation with the person-category approach, which is disaggregated and behaviour orientated. Within the scope of the NRW-FVU it was not possible to create this basis of data. Hence a validation of the models was not possible.

The calculation of the traffic generated in a specific area was done in the following way:

$$F_{A,i,p}^{(k)} = X_{A,i}^{(k)} \cdot MR_{A,g,p}^{(k)}$$

F stands for the trips of zone **i** to be looked at (A = active side, p = trip purpose, k = person category, g = space category), **X** stands for the members of the respective person-category and for the zone of the active side (structure size of the passive side) and **MR** stands for the respective trip rate. The calculated traffic volume is the daily traffic volume. To calculate the emerging traffic during a specific time, it has to be multiplied it with the particular time rate. This time rate can be drawn from surveys. In Fig. 9 the data of departure times are illustrated. Experience shows that all people asked can neither precisely tell their departure nor their arrival times. Thus, in the presented example only 8 % of all data are precise data given in minute intervalls. 81 % of all those asked give only an approximate time (¼ hour) of departure and arrival.

Next to the preciseness of the data, the peculiarity of the pattern of time is dependent on the spatial position of the trip starting places with regard to the places of activity and the respective trip purposes. Looking at Fig. 9, it becomes obvious that the frequencies of the time of departure differ between city and suburban areas. Shorter distances and therefore less travel time are the causes for this phenomenon. Fig. 10 presents travel times in their spatial distribution all over the city (in this case the example is taken from the United States).

The central problem within this phase of the calculation of demand must be seen in the time rates or in the necessary energy to determine them in a way of spatial distinction. Concerning the concrete example of application, namely Wuppertal, the time rates were elaborated from a minor survey of the city and supported by data of a survey taken from Mühlheim as well as from German national travel survey – data (KONTIV – data).

The inland traffic within the zones is usually not adopted into further steps of calculation. Route choice and traffic assignment are only means of calculating the traffic to and from zones. The splitting-off of the inland traffic is realised via splitting-off rates.

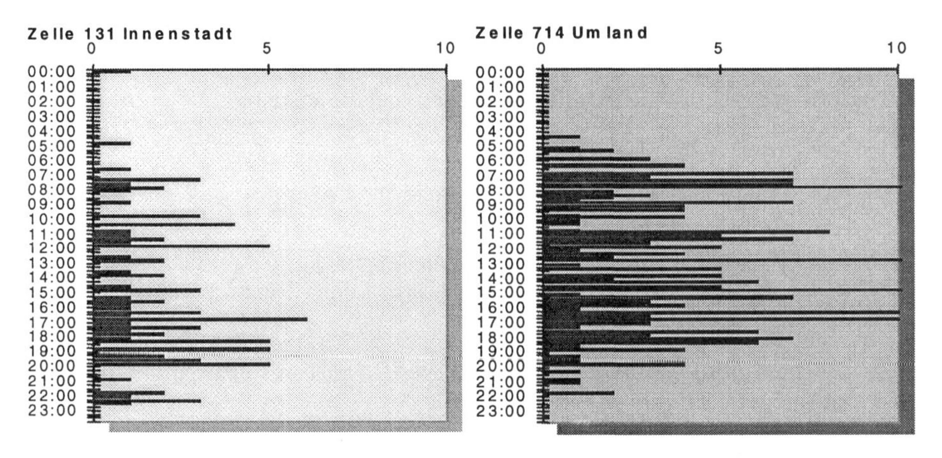

Figure 9: Statement of departure time (source: questionnaire Mülheim/Ruhr; HHS-Harloff Hensel Stadtplanung).

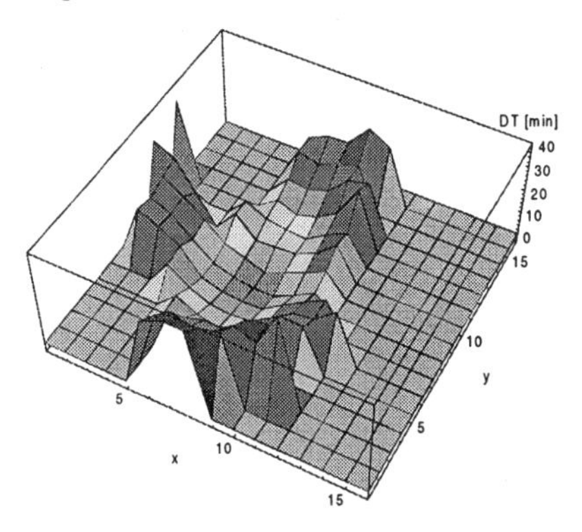

Figure 10: Distribution of the travel time of an urban area (source: P. Wagner, non published paper within the framework of the NRW-FVU).

The algorithm was expanded by the estamination/calculation of the traffic assignment for any time interval, provided that the time rates are available. The time rates are determined by the pattern over time for different trip purposes.

The result of this step is emerging home-based trips Qi (O) and Zj (D) for 8 trip-purpose-person-categories and the different time intervals. These emerging values are subdivided into MIV-bound, public transport-bound and into a facultative choice of transport modes.

Traffic Distribution

The task of the traffic distribution is the assignment of the trip's cause and aim and therefore the construction of O-D matrixes. The available gravity approach was only slightly modified for the application on Wuppertal. Especially with regard to a temporal calculation, traffic distribution should not be neglected methodologically. A basic size is the resistance matrix. Resistance is given for every route within the network. The stronger the resistance (mostly the travel time) between two places of activity, the less trips are put on this relation. Even the competitive situation of the places of activity is taken into account by this resistance. With the help of route choice and traffic assignment the resistance is calculated and would therefore also be available for different time intervals in dynamic approaches.

It is definitely reasonable to present different resistance values (for the calculation of traffic distribution in a gravity approach) for every point in time. Nevertheless, these resistance should be supplemented by "constraints", basic conditions as for instance opening and closing hours of the diverse destinations (institutions). The feeding in of resistance taken from a dynamic route choice and traffic assignment or from a micro-simulation is possible, but for the illustration of the decision what kind of goal has to be chosen, combinations of the expected travel time (resistance) and the possible arrival times are more suitable.

The result of this step are O-D matrices for the MIV, the public transport and the facultative choice for every time interval.

Modal Split II (Traffic Distribution)

In this phase the facultative collective is subdivided into public transport-user and MIV-user. A probit-model is used. An essential part is again the resistance between cause and aim, mostly in form of travel times (sometimes also in form of charges). The development of dynamic methods for traffic assignment and route choice offers the opportunity to calculate the travel times for every point in time so that they can be used in the Modal Split. In contrast to the traffic distribution, the travel time is here a decisive criteria for the illustration of the decision. For the application of the example, first of all, we started from temporal invariable travel times.

Dynamic Route Choice and Traffic Assignment

This phase is not the main part of the work, but the available model is used to calculate the traffic flow in accordance to the CA so that methodological distinctions can be investigated. With DRUM it is possible to calculate routes and traffic flows iteratively for any time intervals. Besides, this step is necessary for a plausibility check of the calculation results by means of the counting values.

Results

For the Wuppertal example, different time intervals (day, hour, one hour, ½ hour, ¼ hour) were calculated. The whole city was divided into 402 zones. The road network consists of 11888 routes and 5600 knots. The model reckons with 12 person-categories, 3 trip purposes and 4 space categories. As a result of this, O-D matrixes for MIV for every time interval are at hand.

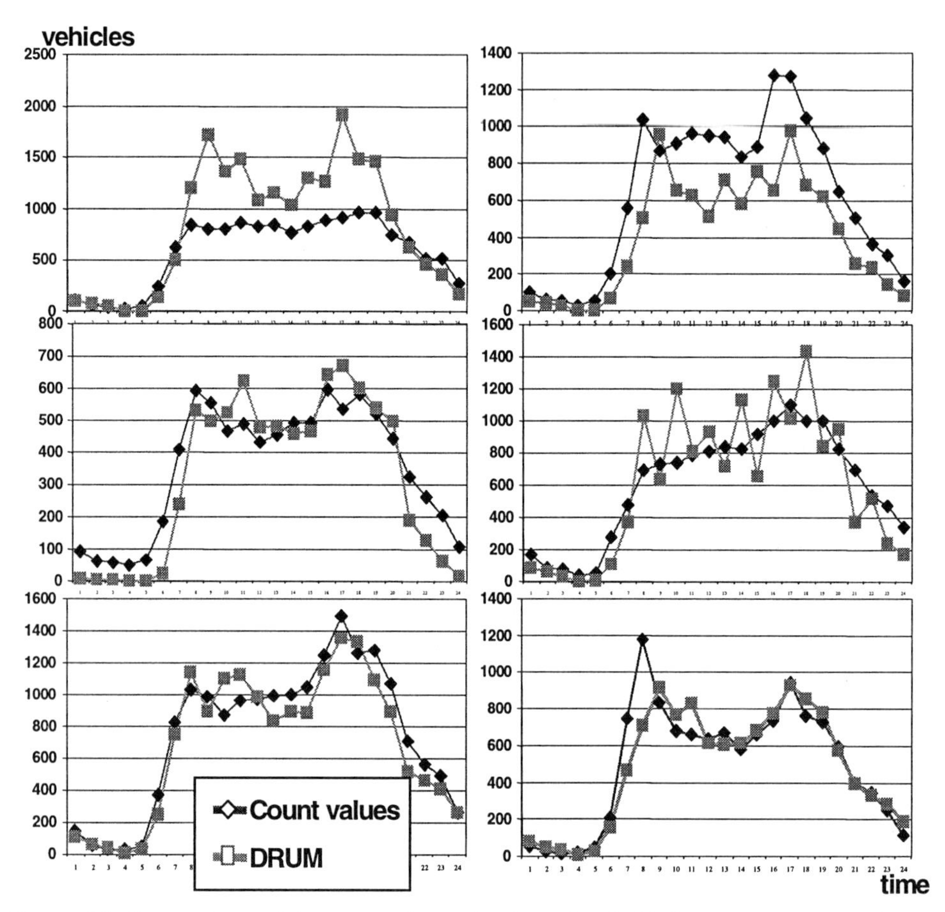

Figure 11: Comparison of the counted and simulated loads (daily pattern over time) at chosen census-points in the urban district of Wuppertal.

As a result of the data which were available as a basis of calculation, neither an official verification nor a calibration of the models can be carried out. Besides, there is a lack of an additional basis of calibration in form of an extensive traffic census. Thus, a rough plausibility check can only proof that the models do not have any serious errors. Therefore, the primary aim is to be seen in opposing the two basic methods concerning the calculation of the route flow, DRUM and CA, and in interpreting and judging the respective results of the calculation by comparison. It is intended to present the results at the workshop.

If we compare the daily patterns over time of counting values and simulation results (first of all DRUM) at census-points within the urban district it becomes clear that even at this point the general use is possible. The patterns over time are well recreated with fair to middling quality at most of the cross-sections which are at hand. Because of the data and the fact that this model has only be used in a "fairly rough state", in only one single calculation (missing calibration / official verification), the results are satisfying. At some census-points the dynamic assignment algorithm shows intolerable oscillations around the real value. These are caused by the interplay of the used speed-flow-pattern over time (speed-flow-curves) and the method of an iterative load elaboration. The speed-flow-curves were formed by present curves corresponding to more extended time intervals by an increase of level, since the short-term efficiency cn be higher the capacity middled over the day. Nevertheless, the limits of capacity are reached during the load's elaboration, and because of the high general load within the network, loads "fluctuate" between alternative sections in succession of the iteration (Fig. 11, diagram in the mid-right). On the basis of this fact, a factor of attenuation was introduced into the algorithm (look Beckmann, Wulfhorst, and others 1997). This version of the algorithm was not used in the present application. Fig. 11 presents chosen comparisons of daily patterns over time. Apart from the very well recreated patterns over time, some are over-, others underestimated. A system is not visible, but the results are highly dependent on the zone's size and in connection to this, on the place of the section the zone connector feeds in. Generally speaking we can say that the results are getting worse within a place at highly loaded cross-sections and at highly aggregated points the zone connector feeds in.

The emerging amount of data during the calculation makes a plausibility check more difficult. This is the reason why the results have been transferred into a geo information system (GIS) and why they have been visualised in a movie-clip, especially for the illustration of the temporal courses (look: http://www.rwth-aachen.de/isb/nrw-fvu.html). Fig. 12 presents one picture of this movie-clip.

Within the scope of the work done by the ISB in NRW-FVU, the use of a GIS is to be seen as an essential part because of its amount of data and the necessity to present (to visualise) results adequately. Thus, with the help of the GIS, link geometry inclusive some of the attributes taken from the ATKIS data (administrative-topographic-cartographic-information-system) could be taken over from Wuppertal. The difficulties concerning acquisition, transmission and modification of data lead to further reflections on the use of GIS-systems in transport planning [12]. The reasonable use of an automatic creation of a calculation network and an automatic standardisation of the routes resulting from the available attributes in ATKIS can be understood as an example. Here, first approaches have been designed and used [13].

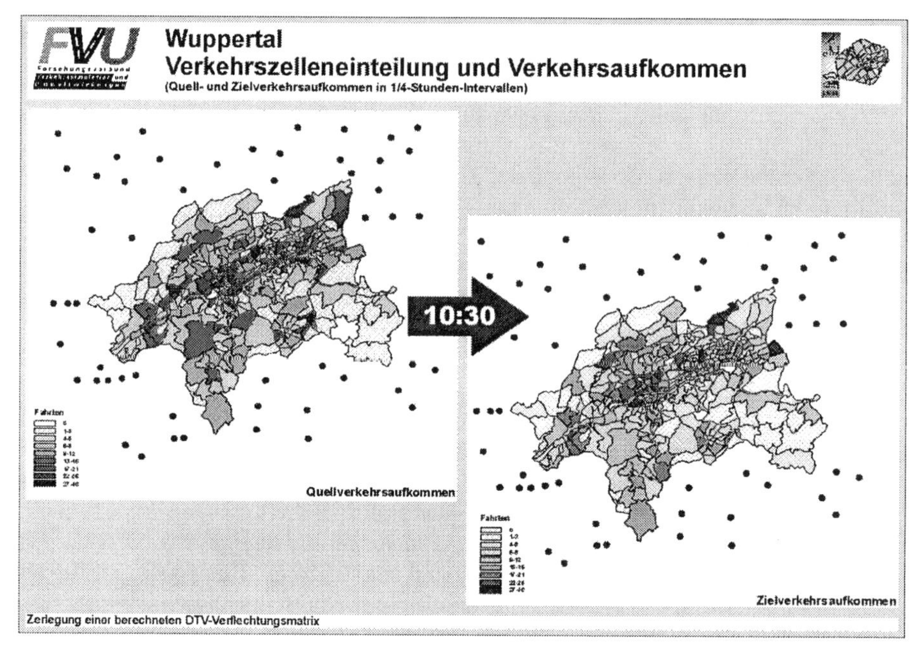

Figure 12: Screenshots of a movie taken on one day of the temporary traffic generated in the specific area of Wuppertal.

4 Use and Methods of Temporal Traffic Simulation – Perspective

In the given examples concerning possible application of the motorway network as well as of the urban road network of Wuppertal, limits of the use of present – partly also typical - methodological approaches of temporal traffic simulation become obvious. For a not insignificant part the present limits result also from nonconformity and differentiation as well as from availability of the basic data.

Among the possible aims, such as the amelioration of the system's management of road infrastructure, road network and road systems, or the creation of "stable" foundations for the On-Line-Management of traffic streams, or the increase of efficiency and the compability of the measurement of strategic planning, the necessity of temporal detailed traffic simulations remains essential. With the help of the already present examinations of the FVU and the experiences already made, it becomes also obvious that promising approaches of simulation (can) demand a coupling of different methods or models.

For an extensive kind of traffic within the scope of the federal motorway (and perhaps also within the scope of federal roads), an obvious amelioration of the quality of illustration seems to be reachable by the fact that

- regional and local traffic, which is handled on federal long-distance roads, and

- "national traffic" as well as traffic of an unique kind (e.g. holiday traffic, short holiday traffic, week-end traffic or long-distance leisure travel)

are treated in a special way during the elaboration of temporal O-D matrixes.

Even in the future, the application of methods for a temporal disaggregation of matrixes, as they have been presented in this text, or the application of similar algorithms seem to be advisable and moderate with regard to the national as well as the unique traffic.

First of all we can say that the reason for this is the fact that

- a temporal distribution for a traffic of an extensive kind , that is to say, long-distance traffic in a "narrower" sense, can only be estimated with a high amount of uncertainty
- there remain special problems concerning the modelling estimation for traffics of a "unique" kind (transport demand, trip distribution, etc.)

Modelling approaches which are stronger oriented on the presentation of the cause of traffic, are to be used for the regional traffic – at this time, especially within the field of passenger transport, but in the future also in the field of freight transport. In this context, approaches concerning person-categories, as for instance for the urban example, Wuppertal, can be utilised. Besides these approaches can be extended or replaced by modelling approaches which base on illustrations of activity chains.

Concerning small scales of consideration of the urban and regional traffic, temporal approaches of simulation seem, in the long run, to be more promising than other approaches, if the models base on the illustration of the person´s activity chains. These modelling approaches can present to every single simulated traveller his own "time information". But difficulties caused by these modelling approaches, as for instance the classification of the trip's aim ("traffic distribution"), the control of the terminating traffic generated in a specific area concerning its limits of capacity as well as the illustration of the specific choice of transport mode are not to be underestimated.

Even with regard to a demand-responsive modelling referring to only one single person (which is probably not a hardware problem nowadays), characteristic features of behaviour have to be at hand in frequency distributions as a data basis. The characteristic features of behaviour can be allocated stochastically. Since the parts within the groups of different characteristic peculiarities can become pretty small, the insecurity of the illustration close to reality can increase. In this way, the critical importance of the available data basis concerning road behaviour, but also concerning offers of traffic networks, basically remain.

In the long run, especially the chances on the On-Line-Application level by coupling of dynamic demand-responsive modelling and the micro-simulation are to be used.

References

1. SERWILL, D.; 1994, „DRUM - Modellkonzept zur dynamischen Routensuche und Umlegung"; Berichte des Instituts für Stadtbauwesen der RWTH Aachen, B43, 1994.

2. Beckmann, K.J.; Wulfhorst, G.; u.a.; 1997, „Verifizierung des Modellkonzeptes DRUM", Schlußbericht zum DFG-Forschungsauftrag SE 818/1-1, Aachen, 1997.

3. Brilon, W. and Wu, N.; 1997, „Calibration and validation of Cellular Automaton. Technical Report. Institut for traffic engineering, Ruhr-Universität Bochum, 1997.

4. Nagel, K.; Schreckenberg, M.; 1992, A cellular automaton model for freeway traffic. J. Phys. / France, 2:2221-2229.

5. Wagner, P.; 1995, Traffic simulations using cellular automata: Comparison with reality. in: D.E. Wolf, M. Schreckenberg, A. Bachem (Edts), Proceedings of Workshop in Traffic and Granular Flow, World Scientific, Singapore 1995.

6. Wermuth, M.; Gudehus, V.; Zumkeller, D.; Meyburg, A.H.; 1990,„Modelltypen zur Verkehrsprognose und ihre Einsatzfelder"; Schlußbericht FE 77042/85 im Auftrag des Bundesministers für Verkehr; 1990.

7. Bundesminister für Verkehr; 1996, „Entwicklung von Verfahren zur großräumigen Prognose der Verkehrsentwicklung und Folgerungen für den Datenaustausch von Verkehrsrechnerzentralen" Teil I: Prognoseverfahren, Forschung Straßenbau und Verkehrstechnik Heft 727, Bonn-Bad Godesberg, 1996.

8. Bundesminister für Verkehr; 1993, „Typisierung von Dauerlinien stündlicher Verkehrsstärken des Kfz-Verkehrs nach Fahrtzweckgruppen und langfristige Veränderungen", Forschung Straßenbau und Verkehrstechnik Heft 657, Bonn-Bad Godesberg, 1993.

9. FGSV - Forschungsgesellschaft für Straßen- und Verkehrswesen; 1995, „Schätzung von Verkehrsbeziehungen mit Hilfe von Querschnittszählungen", Köln, 1995.

10. BASt - Bundesanstalt für Straßenwesen verschiedene Hefte zur Verkehrsentwicklung auf Bundesfernstraßen, Berichte der Bundesanstalt für Straßenwesen, Verkehrstechnik, Bergisch Gladbach.

11. Gawron, C.; Rindsfüser, G.; 1998, unveröffentlichte Arbeiten im Rahmen des NRW-FVU zur Simulation des Autobahnverkehrs mit CA und DRUM.

12. von der Ruhren, S.; 1998, Nutzungs-/ Anwendungsmöglichkeiten von Geoinformationssystemen im Aufgabenfeld der Verkehrsplanung, Diplomarbeit am Lehrstuhl für Stadtbauwesen der RWTH Aachen, 1998.

13. Rindsfüser, G.; (1998), Automatisierte Typisierung von Verkehrsnetzen für die Verkehrsplanung aus Geobasisdaten. In: M. Schrenk (Hrsg.), Computergestützte Raumplanung. Beiträge zum Symposion CORP'98, Band 1, Wien, S. 177-182.

The Development of a Unified Modeling Framework for the Household Activity-Travel Scheduling Process

S.T. Doherty[1] and K.W. Axhausen[2]

[1]Department of Urban and Regional Planning, University of Laval, Quebec City, Quebec, Canada, G1K 7P4

[2]Institut für Straßenbau und Verkehrsplanung, Leopold-Franzens-Universität, Technikerstr. 13, 6020 Innsbruck, Austria

The goal of this paper is to propose a new approach to activity-travel schedule modeling that provides a unifying framework for past research in different areas. This approach is based on empirical evidence gathered using a Computerized Household Activity SchEduling (CHASE) survey. The survey provided a means to examine the underlying scheduling behavior of household over a one week period as it occurs in reality. Results show that a clear distinction can be made between routine scheduling decisions that are pre-planned before the week commences, and the more short-term, impulsive, opportunistic decisions made as the schedule is executed during the week. This distinction allows one to conceptualize the modeling task as a multi-stage process, wherein routine planning is approached with existing optimization models (assuming that routine activities are the result of a long-term thought and experimentation process) followed by a more sub-optimal rule-based simulation model to replicate the decisions process during the week within the constructs of the optimal routine plan. Such a model is proposed in this paper as a long term development, and would rely on the type of data provided by new data collection techniques such as CHASE. Operationalization of the model as an event-oriented simulation is proposed. Various components of the model are explored in detail, and discussed within context of existing models.

1 Introduction

Activity-based approaches to travel analysis have made extensive contributions to the understanding of travel behavior and the likely impacts of social and policy changes. The 1970s saw the initial development of the theoretical framework for viewing travel as part of an activity-based framework, whereas the 1980s saw major advances in methodology, particularly model development, as well as initial application of models to policy analysis. The early nineties have seen important contributions to the development of modeling techniques which have become more operational but require further validation and unification before they are fully applied in practice.

The major impetus for this research is a need for improved models to assess the impact of travel demand management policies that have emerged as the focus of transportation policy. These needs have led to the development of models of specific aspects of activities, and more recently to the modeling of entire activity-travel patterns or "schedules".

Models of Activity-Travel Schedules

Activity-based models have sought to replicate one or more of the four basic dimensions of activities: i) activity choice; ii) duration (over what time periods); iii) location; and iv) sequencing (i.e. when they take place, in what order, and with what frequency). Many of the first models sought only to replicate certain dimensions, such as time allocation to activities. Later models incorporate all dimensions within their model in an attempt to replicate the way individuals and households arrive at a total activity pattern in time and space. These are often grouped under the heading of activity "schedule" models, since a schedule of activities embodies all four of these dimensions.

Many different categorizations of activity scheduling models have been made and discussed in the literature (for more thorough reviews see [1, 2, 3]). Two important dimensions of these models are 1) whether the modeling approaches are static in nature (i.e. estimate activity patterns *simultaneously* in one step) or *sequential* (i.e. step-by-step structure), and 2) whether they adopt an *econometric* (e.g. utility maximization and optimization) or *rule-based* psychological approach (i.e. sub-optimal, satisficing style rules) to model development. Distinctions are also made between theoretical versus operational models, the latter being exclusively based on observed data.

Previous modeling efforts have relied mainly on the traditional utility maximization framework to replicate specific aspects of the scheduling process in limited combination or to capture the static choice of an entire daily activity schedule. Three basic approaches have been adopted. The first is to model the sequential choice of activities and locations to add to a sequence/pattern of activities and travel (e.g. [4, 5], and [6]). Conventional logit type models are used to predict the choice(s) conditional on the characteristics of activities/locations that proceed and follow the given choice. A second approach focuses on the simultaneous choice among a set of activity-travel sequences/patterns, rather than on their sequential construction (e.g. CARLA [7], STARCHILD [8, 9], and [10]). A third approach that represents a mix of the sequential choice of pre-defined patterns is the tour based nested logit model developed by Bowman et al. [2, 11, 12].

However, the behavioral validity of the utility maximization framework as a description of how people actually make decisions has continuously been questioned [13]. Criticisms focus on assumptions regarding full information and the capacity of individuals to determine optimal solutions. Specifically, the main criticism of simultaneous schedule choice models is the assumption that people choose amongst a large set of patterns to maximize their utility, which is viewed as unrealistic since people are rarely aware of all possible patterns available to them.

In the case of *tour* based models, the behavioral assumption is that the activity scheduling decision structure consists of a series of choices that can be described by discrete choice models. However, simply adding more nests to the utility maximization model to account for the complexity of scheduling is insufficient on behavioral as well computational grounds. For instance, Bowman and Ben-Akiva's [2] attempt to incorporating the time dimension led to the addition of another nested level to their model. Not only does this involve a behavioral assumption about how timing decisions are made, but computational limits meant that only four time periods could be included for choice, which is clearly inadequate for policy analysis. Extension to more precise time periods would cause serious problems in a nested logit system.

Several models appear to be well conceived from a theoretical standpoint, as was the case with STARCHILD and Bowman and Ben-Akiva's models, but their operationalization suffered from a limited focus on the use of utility maximization discrete choice techniques. Insights from cognitive psychology about how people perform complex scheduling tasks suggests that people apply a large range of heuristics and strategies when faced with such tasks [14, 15]. Gärling et al. [16, p. 356] argues that an even more serious issue relates to the tendency of traditional models to be confined to specifying what factors affect the final choice of *pattern* whereas the *process* resulting in the choice "is largely left unspecified."

In response to these criticisms, recent modeling efforts have attempted to more explicitly replicate the *sequencing* of decisions made during the scheduling process, under alternative "rule-based" behavioral structures. These include SCHEDULER [16, 17], SMASH [18], and models by Lundberg [19], Gärling et al. [20], and Vause [21]. The most advanced operationally of these models is the SMASH model. The model starts with similar inputs to the CARLA and STARCHILD models (a list of activities to schedule along with their attributes), however, an individuals schedule is successively constructed by maximizing utility at each step (add, delete, reschedule) taken to construct the schedule, rather than for the schedule as a whole. Although this model is highly innovative, several key criticisms can be noted. Firstly, the authors note that "the mechanisms of the model allow for the adjustment of the schedule during the travel phase", however, this adjustment is limited only to substitution of activities between the agenda and schedule, ignoring other adjustment possibilities such as changes in activity duration and location. The authors also note that changing travel times and unexpected activity durations effect the utility in a sense that the chance of completing the schedule may vary, but they do not go on to describe how the adjustment of the schedule during execution is incorporated in the model. In this way, the SMASH model is still somewhat limited to the same "pre-travel phase" of scheduling as the STARCHILD model.

Overall, these behavioral limitations may seriously hamper the use of SMASH to assess the impacts of policy that inherently invoke a rescheduling response involving more than just substitution of activities (e.g. tele-communication), or that involve changes in behavior during execution of the schedule (e.g. ITS - Intelligent Transportation Systems). These criticisms may be due in part to the limitations of their interactive computer experiment MAGIC [22], which was limited to investigating activity scheduling behavior in a lab setting essentially ignoring the portion of scheduling decisions made during execution of an individual's schedule. The Ettema findings were also used in general to support the more recent sequential ruled-based models of Gärling et al. [20] and Vause [21], which may explain why they too are somewhat limited in the extent to which they address how scheduling decisions are subsequently modified during execution, and why they struggle with assumptions concerning the sequencing of decisions.

Future Directions

From the literature, it is clear that contributions to the complex field of activity scheduling are widely dispersed. Scheduling models based on existing econometric techniques are severely limited by their behavioral assumptions which limit their applicability to policy assessment. Alternative theories about activity scheduling that have been introduced are quite difficult to operationalize in their full form (e.g. SCHEDULER, STARCHILD). Operational scheduling models that have been developed (e.g. SMASH) have been criticized by practitioners for their complex data requirements and as discussed, are still limited in their ability to capture more complex rescheduling of activities. This scattered nature of the research is reflected in Ettema and Timmermans [1, p. 33] conclusion that

> "From a scientific point of view, it can be argued that to date a considerable body of knowledge exists regarding aspects of activity and travel patterns. At the same time, however, it should be noted that research in this area has been fragmented and that a *unifying framework* which links researches in different areas is still missing. This is probably due to the complexity of the phenomenon and the applied nature of most activity-based research."

What is not made clear in the literature is that observed activity schedules are the result of an unobserved decision "process" involving the planning and execution of activities over time within a household context. Travel behavior researchers are increasingly recognizing the need for in-depth research into the household activity scheduling "process" in order to advance model development. For example, Pas [23, p. 461] noted that "understanding travel and related behavior requires the development of models of the process by which travel and related behavior change" while Polak and Jones [24, p. 2] state clearly that "the degree to which travelers will be able or willing to adjust the timing of their journeys in response to Road Pricing charges will ultimately depend upon the nature of these scheduling processes. The development of improved understanding of these processes and the translation of these understandings into operational modeling techniques is a major research priority." Axhausen and Gärling [3] emphasize in general, that the rescheduling of activities is at the core of many of the changes in travel behavior brought on by recent policy initiatives related to information technology and transportation demand management. Thus, it is becoming ever more important that the development of travel forecasting models capable of assessing these types of emerging policies need to explicitly account for how people would temporally and spatially adjust their travel behavior, which is dependent on an underlying process of activity scheduling.

Linking Activities and Integrated Urban Model

Growing environmental concerns, and the awareness that long term reductions in emissions requires transport as well as land-use policies, has renewed interest in integrated models of Land-use, Transportation, and the Environment (LTE). Wegener [25, 26] stresses that future LTE models need to respond to a new generation of activity-based travel models that require more detailed information on household characteristics and activity locations. For the most part however, LTE modeling has continued in a business-as-usual fashion, focusing on single-purpose trips and the integration of traditional travel demand models. One state-of-the-art microsimulation model of (L)TE that does incorporate an activity focus is the TRANSIMS model [27, 28]. The "Household and Commercial Activity Disaggregation Activity Demand" sub-module is designed to be probabilistic in nature, in that for a given set of households a distribution of activities and their attributes are produced. Attributes include activity importance, the activity duration, activity location (for mandatory activities) and a time interval during which the activity can be performed. The mechanisms used to actually *schedule* the list of activities is not described in either paper, but was identified as a major "question mark" for future development at a recent conference [29].

A conceptual framework of how an activity scheduler would contribute to an LTE model is presented in Figure 1 (italics in the text represent components in the figure). For a similar approach see for example Axhausen and Goodwin [30]. The upper portion of the model focuses on long term *Land-use* and demographic processes, including *Household Demographics*, *Residential Location*, *Employment Location*, *Vehicle Ownership*, and *Firm Location*, and the *Road/transit Network*. Each of these sub-modules are currently being developed within the micro-simulation platform of the ILUTE model currently being developed in Canada [31]. It is proposed that several of these sub-modules will input information to the *Household Activity Agenda* and/or *Household Activity Scheduler,* at specified intervals or as events unfold in the micro-simulation to support creation of new travel demands.

The *Household Activity Agenda Simulator* consists of a list of household activities, along with the salient attributes that influence their scheduling, such as their desired frequencies and durations, possible start-end times and location choice sets. Some of these attributes would likely be probabilistically related to individual/household characteristics based on activity diary data (or perhaps, modified travel diary data), in a similar approach to that adopted in the TRANSIMS model. The *locations* of home, work, school/daycare and other mandatory activities could be taken as given from previous sub-module steps. The location choice set of other activities would require a model of an individuals cognitive map, perhaps simulated based on residential and employment location histories. The adaptation of the activity agenda in the long term, such as in the case of learning new activity locations or activity types, would be a necessary component of this sub-model. Simulated changes in the agenda as a result of policy (e.g. shorter working days, longer working hours for females, increased telecommuting) would be reflected in changing attributes of certain activities that effect scheduling patterns and resultant travel patterns. Spatial, temporal, coupling, institutional, household resource, and transportation related constraints need also be imbedded in the structure of the household activity agenda. For instance, a household constraint that parents be at home at a certain hour to care for their children would be represented as a pre-planned activity with highly fixed time and location.

The *Household Activity Scheduler* would take the agenda of household activities and model the steps/process by which the activities are sequenced in time and space. Such a model is the focus of the remainder of this paper. The output from the *Household Activity Scheduler* would include the *Travel Demands* of each household member by time of day. This would feed into a *Traffic Flow Model* in order to generate network flows and updated travel times due to congestion. The updated travel times can be used to feedback to the Household Activity Scheduler in the random events that could result in further scheduling modifications. The activity scheduler could also be used to feed information to a residential choice model, in the form of the variables that indicate the potential utility of activity patterns associated with a set of residential choices for a given household. Practically, the activity patterns that could be feasibly evaluated by a household in a new location would be restricted to high priority or "routinized" activities. This would expand upon traditional residential choice models which use only "work trip" accessibility as a variable in the model (e.g. [32]) and could provide especially useful for certain population segments that are relatively insensitive to the work trip accessibility (e.g. telecommuters). Keeping track of residential mobility in a microsimulation may also play a complementary role in the construction of activity schedules, allowing one to consider history dependent variables that effect activity schedules (e.g. length of residence). Exactly how the linkages are made, and at what time scale are additional issues that must be resolved.

2 The Weekly Household Activity Scheduling "Process" Model

Data Collection

Despite the need, very few data collection efforts have targeted the underlying activity scheduling process. Exceptions include Hayes-Roth and Hayes-Roth [15] who used a "think aloud protocol" to investigate the kinds of behavior exhibited when people are posed with a series of errands to perform, and Ettema et al. [22] who used an interactive computer experiment to identify the types of steps people used to construct a one-day schedule. The CHASE (Computerized Household Activity SchEduling survey) survey developed by Doherty and Miller [33] goes beyond these methods by providing a means to observe the scheduling process as it occurs in reality in a household setting over a multi-day period. In this way it is able to capture both routine and complex scheduling processes as well as observe those scheduling decisions made during the actual execution of the schedule.

The CHASE program is designed to track the sequence of steps taken by individuals in a household to add and subsequently modify/delete activities from a household "agenda" to form weekly activity schedules. An upfront interview is used to establish a household's activity agenda which consists of a full list of activities potentially performed by household members, along with their attributes. This information is entered by an interviewer into computerized "forms" linked to a database file that the CHASE program can access in order to display the information back to the user in choice situations. Users are basically instructed to login daily to the program for a week long period (starting Sunday), and continuously add, modify, and delete activities to an ongoing display of their weekly schedule (Monday-following Sunday), not unlike a typically dayplanner. Aside from these basic scheduling options, the program automatically prompts the user for all additional information. The result is a highly detailed trace of the scheduling decisions adults in a household. Doherty and Miller [33] show that the program has a relatively low respondent burden and minimizes fatigue effects commonly associate with multi-day surveys. This approach goes a long way towards solving the data collection problem highlighted by Bowman and Ben-Akiva [2] that simulation models of activity scheduling require "very complex surveys for model estimation" wherein "respondents must step through the entire schedule building process." CHASE data can also provide considerable support for other sequential decisions process models, such as those proposed by Gärling et al. [20] and Vause [21], that have lacked direct empirical support.

CHASE data from a sample of 40 households (55 adults) in Hamilton Ontario is used in this paper to support the proposed model. The households represented a roughly equal mix of married couples, married couples with children, and single person households. The majority of households were located within two kilometers of the McMaster University campus, which is situated at the very tip of the western end of Lake Ontario (Hamilton region population: ~600,000).

Model Structure

The following conceptual model is presented as a means to describe how past research can be brought together into a unified modeling framework, and to lay the groundwork for future operationalization of the model. Note that italicized terms in the text refer to the model components as depicted in Fig. 2 and Fig. 3. In general, the model attempts to dynamically replicate the scheduling process as it occurs over time through the use of various modeling constructs and decisions rules. It begins by taking an individual's *Household Agenda* of activities, and establishes a set of *Routine Activities* and a *skeleton schedule* for the individual for the week, via an optimization model. This is followed by a *Weekly Scheduling Process Model* that replicates the scheduling decisions (additions, modifications, deletions) made by individual during the execution of their schedule during the course of the week. An activity priority function combined with various decision rules are key ingredients in the simulation model. Each of these aspects of the model are described in more detail in the following sections.

The Household Activity Agenda

On a fundamental level, activity scheduling reflects personal and household related basic human needs constrained in time, capability, and in space by the urban environment. These needs can be viewed as manifested in a household's activity *agenda* which represents the initial input to the model as shown in Fig. 2. The agenda consists of a list of uniquely defined activities that a household could potentially perform. Each activity on the agenda is viewed as having a unique set of (perceived) attributes that affect their scheduling, including duration (min, max, mean), frequency, earliest and latest end times, mandatory/optionally involved persons, costs, perceived locations, etc. These rather flexible parameters are used to determine the exact start/end times, location, etc. of the activity once scheduled, as shown in Fig. 3 (*Refine A Choice*). What is key to the success of the scheduling model is not the activity types as defined by traditional means (e.g. work, school, shopping, mandatory, discretionary etc.), but rather that their salient attributes are unique, giving the model the ability to address any number of individuals/household types. Although the derivation of household activity agendas are of considerable interest on their own, they are taken as exogenous to the process of scheduling in the short term (see also Figure 1 and related discussion).

Routine Weekly Activity "Skeleton"

Empirical evidence derived from the CHASE survey shows that households begin the week with a planned set of *routine weekly activities.* On average, 45% of weekday and 20% of weekend activities were pre-planned on the First Sunday of scheduling (remembering that users began scheduling on a Sunday for the activities that take place Monday to the following Sunday). This represents a total average of 34 activities per adult pre-planned on the first Sunday. Of the decisions, a full 70% were part of multi-day entries (the activity was added on 2 or more days simultaneously), with 80% of these consisting of entries across 4+ days. Comparatively, on Monday, only 21% additions were part of multi-day entries, followed by 2%, 6% and no more than 1% on remaining days of the week. Such repetitive entries are indicative of highly routine activities. Other characteristics, such as longer durations (double those of other planned activities) and a higher degree of spatial-temporal fixity, differentiate these routine activities.

Further empirical analysis using an appropriate discriminant analysis technique will be performed to further differentiate these types of activities from other activities on their agenda based on their key attributes, including duration, frequency, and indicator variables of temporal and spatial fixitivity. The resulting D*iscriminant Function* would be used to establish the R*outine Activity Subset*, as displayed in Fig. 3. It is reasonable to assume that these routinized activities pre-planned before the week starts are the result of a long-term thinking and experimentation process, and thus represent an "optimized" pattern or "skeletal" basis around which other scheduling decisions are made during the week. Given this, it is reasonable to assume that an *optimization model* would be appropriate to derive the *Pre-week skeleton schedules*. This model would use the discriminated activities as input, which represent a much more limited choice set of activities that are more amendable to the assumptions that underlie these models. These techniques include those that start by generating all possible feasible combinations of skeleton structures and choosing the most optimal of the set (e.g. CARLA, STARCHILD). The notion of adopting existing models for this purposes is discussed further in the concluding section.

Weekly Scheduling Process

Results from the CHASE survey show that after the first Sunday, a more active, opportunistic, and impulsive mix of decisions follows. On average, adults make about 8 additions, 2 modifications, and 1 deletion per day during the execution of their schedule over the course of the week, which include an average of 12.4 activities and 4.9 trips per adult per day. These scheduling decisions are made on a variety of time horizons. Outside of the routine activity additions made on the first Sunday (38% overall), a substantial proportion of additions are scheduled *impulsively* just before execution (28% overall), on the *same day* (20% overall), or are *pre-planned* one or more days in advance during the week (15% overall). When *pre-planning* during the week, adults were found to reach out beyond one day 38% of the time to make an addition, in an opportunistic fashion. The distribution of time horizons for modifications and deletions differed, as more impulsive modifications occurred (62%), while more deletions are made the same day (38%), reflecting more forethought for deletions compared to modifications.

This evidence strongly suggests that activity scheduling is a dual process of routine optimal planning, followed by a more dynamic process of continued pre-planning, revision, impulsive, and opportunistic decisions made over the course of the week within the bounds of an optimized skeleton structure. Given the goal is to develop a behaviorally sound model, a modeling structure is needed that can simulate this latter process. The *Weekly Scheduling Process Model* shown in Fig. 2 (and elaborated upon in Fig. 3) attempts to fill this gap.

Given that routine scheduling decisions are conveniently made in advance, it follows that they would form an input to this simulation model, as does the agenda of household activities. At this point however, some of the activities on the agenda will already have been placed on their scheduled, and while they always remain on the agenda, their "priority" for subsequent addition and/or modification will change over time in response to changes in the schedule (see the following section for details). The notion of a changing or "momentaneous" priority of activities is viewed as the driving force behind the variety of decisions made during scheduling (for similar ideas see [34] or [35]). The scheduling process simulation depicted in more detail in Fig. 3 will incorporate a mix of empirically derived "priority" functions and decision "rules" that serve to sequentially simulate the series of additions, modifications and deletions taken to construct the activity schedules of individuals in a household over time.

The simulation begins with a set (or stack) of simulated individuals who share an activity agenda with other household members. Individuals are visited and re-visited in sequence at the beginning and end of each scheduled activity or empty time window on their schedule. These visits are ordered in time along with all other individuals such that everyone's schedule is constructed simultaneously. Once an individual is visited, they are faced with a *choice to add* activities anywhere in their schedule, including in the immediate time slot (an impulsive decision) or at a later time slot (on the same day, or one or more days in advance). The priority of activities on the household agenda at that particular moment in time determines what, if any, activity will be added (the priority function is described in the next section). Once an activity is deemed high priority enough to schedule, *feasible windows* of opportunity are defined and one is chosen. Random events may also be generated at the same time that activities are being added to the schedule. Random events include both random changes in activity duration or travel times, or the generation of unexpected or emergency activities (accidents, surprise visits). These latter activities would automatically be assigned the highest priority for scheduling.

Only after this point are refined choices in the activity made, including decisions about travel if needed (mode, route choice, etc.), exact start/end times, exact location, and involved persons. Some of these choices may already be fixed (e.g. location is fixed to at home). Other choices will be simplified, if not limited in their choice set, given the fact that the individual may already be placed in a given spatial-temporal situation that constraints their choice. For instance, if the person is already at certain location outside the home, with a car and with their spouse, and is making an impulsive decision, the mode and involved person choices are somewhat fixed, whereas the location choice set can be simplified given the proximity of perceived locations in the area. When faced with more flexibility in scheduling, factors such as how many other high priority activities are on the agenda, and the desired attributes of the current activity (e.g. desired duration) will effect the refined choices.

Once the final refinements are made, the activity is placed on the schedule of the individual, and a decision is made whether to continue scheduling at the time. This will depend on the number of high priority activities on the individual schedule at the moment. If yes, the process repeats itself, otherwise the individual is placed back in the stack of all individuals, which is ordered in terms of when each individual is visited next.

Conflicts that arise due to random events, the need for more time to accommodate a high priority activity, or cases where activities may be extended to fill time, are handled by the *Modify and Conflict Resolver*. Results from the CHASE survey indicate that the most common modification is to the start or end time of an activity, representing 73% of all modifications. Changes in involved persons (8%), activity type (7%), location (6%), travel time (4%), and mode (2%) were recorded less frequently.

This suggests that people are most often responding to time pressures when they modify activities. People also tend to modify more than one attribute of activities to accommodate scheduling changes, as more that one half the 1241 recorded modifications involved a change to two or more attributes. The *Modify and Conflict Resolve* is intended to take the previously scheduled activities and determines those which have the highest priority for modification. A set of possible modifications is determined, and a choice is made as to which ones to implement and their extent. If the (set of) modification(s) does not meet the requirements of solving a conflict, then the *deletion of an activity* is considered. The procedure for deletion is similar to modification, except that the activity attempting to be scheduled is compared directly to the revised priority of the activity chosen for potential deletion. If none of the deletions is justifiable, then the model reverts back to the beginning, and the originating activity is left unscheduled. If an activity is deleted, control reverts back to the assessment of window feasibility.

Although the scheduling simulation proceeds in a sequential fashion, without directly involving the optimization of the schedule as a whole (apart from the optimization already achieved via routine scheduling), a degree of sub-optimization is achieved by revisiting previous activities for modification. This leads to more optimized locations, durations, mode choices, etc. that minimize travel time or durations via the *Modify and Conflict Resolver*. However, this occurs only in the event that other activities of high priority need to be scheduled within limited time windows. Behaviorally, this reflects the notion that people consider optimizing their behavior only when and where needed and/or possible.

Priority Function and Decision Rules

Although activity "priority" has been proposed as the determining factor in the choice of activities to schedule in previous models (e.g. SCHEDULER [16-36]), it has remained a difficult attribute to operationalize. Asking people to assess the priority of a list of activities is difficult not only because of a definition problem, but because the priority of an activity depends on the situation at hand. Any one static assessment of the priority of activities will be inadequate to deal with all possible situations that arise during the scheduling process.

To meet this challenge, the priority of activities on the household agenda should be modeled as a function of the attributes on the activity on the agenda (duration, frequency, etc.), as well as the attributes of the activity relative to the scheduling state of all household members at the time of decision making. For instance, history and future dependent variables that account for the likelihood that activities that have been already scheduled, or that have taken place recently relative to their desired frequency, would have lower priorities. The temporal and spatial fixity/flexibility of activities suggested to influence the sequencing of activities [37, 38] and hence their priority, could also be investigated by a combination of activity and scheduling attributes. For instance, the number of perceived locations in the vicinity of an individual's current location could be used as a measure of the spatial fixity, whereas the ratio of minimum activity duration to the difference between the earliest possible start and latest end time could serve as indicators of a temporal fixity. The flexibility of an activity in terms of duration in relation to the maximum size of any feasible time windows (max *W*) on a schedule should presumably influence its priority for scheduling. An appropriate variable for the priority function would then be:

$$\frac{\min d_i}{\max W}.$$

The smaller this is, the higher the priority should be. For activities that represent tasks to be assigned to household members (e.g. shopping), the same variable for other household members could be included to reflect the lower priority for an individual when other household members schedules are relatively more flexible. The high priority of joint activities that reflect household constraints (e.g. chauffeuring) could be reflected in a dummy variable that is set to 1 for activities that have already been scheduled by the other household member. The proximity of perceived locations for the activity in combination with available modes and travel times, should also effect priority. Many other variable are possible to reflect the changing level of priority of activities, and to capture the seemingly complex array of decisions and resulting patterns that result. Separate models need be constructed for the priority of activities for addition, modification or deletion to the schedule. Future estimation of these priority functions will be possible using CHASE data.

Throughout the simulation, decision rules are used to replicate the variety of choices made. In some cases, these rules may be rather simple reflecting practical considerations or straightforward logic, whereas in others, they will reflect more complex decisions structures. For example, the decision rule to determine which *activity to add* from the agenda could be as simple as choosing the one with the highest priority. A more complex decisions rule example would apply to the choice of whether to *Continue Scheduling*. The decision rule could be of the form:

IF [Priority of highest activity] > (α) THEN [*continue*]

where the α threshold value is determined empirically from observed data, based on how much free time is left on an individuals schedule before they stop pre-planning. An alternative would be to base the decision on the sum of priorities of activities on the agenda, replacing the left side of the above equation with the sum of all priorities on the agenda. This would reflect the aggregate amount of pressure the particular person is under to continue scheduling activities. This rule would need be combined with other practical rules that require that scheduling proceed in light of any open time windows, regardless of activity priorities. Other rules would be needed for choosing *Feasible Time windows* (closest in time?, longest?) deciding when modifications/deletions are needed (when priority of an activity is sufficient high relative to scheduled activities to justify modification/deletion to accommodate its scheduling), deciding which modification(s) to make, and the variety of activity choice refinements that need be made. Clearly, much more work is needed on the cognitive behavioral side to improve these rules, however, simplified rules could be used to operationalize the model in the short term.

Operationalizing the Simulation

The scheduling process suggested in the discussion above and summarized in Fig. 3 has to be operationalized as an event-oriented simulation, which is able to model the interactions between persons in time and space [39]. Event-oriented simulations divide all operations of the model and of the entities, here persons, into individual *events,* which encapsulate a particular set of actions and which are executed at a particular moment (simulated time). Each event selects the relevant next event, which needs to be scheduled for the entity concerned and calculates the time when this event is going to be executed.

At this point it is not possible to give a full list of all relevant events, as this list will depend on the amount of functionality envisaged for the initial implementation. Still, the following core will be required (next event to be scheduled):

Agenda construction: constructs the initial agenda for a household. (Skeleton optimization)

Skeleton optimization: selects the routine activities from the agenda and constructs the optimal skeleton schedule. (Preplanning)

Preplanning: For the remaining days of the week the schedule is advanced, i.e. activities added, modified and deleted. (Day of)

Day of: For the remainder of the day the schedule is advanced. (Preplanning or Impulsive)

Impulsive: the next activity is selected by finalizing the *local* schedule by filling the time window immediately in front of the person, implying choice of all aspects of the movement to the chosen location (mode, route/lines, transfer points, preferred parking location/stop, preferred type of parking, acceptable search time). (Day of or Activity start)

Activity start: If the activity is undertaken at the same location, then the following calculations are undertaken:

- The activity duration is finalized by randomly drawing the duration, as a function of the anticipated duration: $\Delta t(\text{duration})$.
- Given the activity duration the occurrence of a random external event requiring scheduling is determined at $\Delta t(\text{random}) < \Delta t(\text{duration})$ including the type of event.
- Schedule next *Impulsive* at $t+\min(\Delta t(\text{random}), \Delta t(\text{duration}))$, which is implicitly the end of the current activity, unless this is the last activity of the day, i.e. going to bed for the night.

If the next activity is elsewhere, schedule *Movement Preparation* at t+min Δt.

Movement Preparation: Based on current information and the prior choices, confirm the route and mode chosen. If the expectation is, that movement can be performed in the time allocated, schedule *Move* at t+min Δt, otherwise schedule *Impulsive* at t+min Δt.

Move: Move down the next (first) link of the current route, calculate travel time as a function of the link usage Δt(link). Revise preferred parking location/stop, if required by actual travel conditions. If outside anticipated time limit for this point, schedule *Impulsive* for t+Δt(link). Otherwise schedule:

- If parked vehicle reached, schedule *Move* at t+Δt(Getting the vehicle started), which depends on the type of vehicle, the type of parking, the size of the group and the purpose of trip.
- If initial stop reached, schedule *Move* for t+Δt(until the next arrival of vehicle of preferred line).
- *Move* at t+Δt(link), if not yet at preferred parking location, stop, destination.
- If at the preferred parking location, schedule *Parking* at t+Δt(link).
- If at preferred stop, schedule *Move* at t+Δt(link) for final walk to destination.
- If destination has been reached (by walking), then schedule *Impulsive* at t+Δt(link).

Parking: If parking of the preferred type is available within the acceptable time frame, calculate Δt(search time) and schedule *Move* at t+Δt(search time) for the final walk to destination or next mode, in case, for example, of P+R or Kiss+Ride. Otherwise, schedule *Impulsive* at t+Δt(wasted time), which depends on the type of parking preferred and the acceptable search time. (It is assumed, that the duration of the parking search can be determined from the number of vehicles travelling on this link and those searching for parking. See [40] for an example and the literature cited there).

This formulation is open to include a whole range of further events and interactions, which might be of interest to a particular context. For example:

- Simulation of telecommunication by including interrupts of activities of others, as part of *Impulsive.*
- Traveler information by adding *Impulsive* scheduling events depending on whether a traveler has received certain information while traversing the current link.

- Simulation of public transport vehicles and their interactions with other traffic and the resulting early/late arrivals or changed waiting times.
- Detailed simulation of traffic control devices, in particular signals, by modeling detector locations and the associated adaptive signal control.

The exact implementation of the event-driven simulation is a question of the available computing resources, but current programming tools, including agent-based languages, are greatly facilitating the task.

3 Discussion

Model Comparison

The conceptual model presented in this paper can be compared to previous sequential activity scheduling process models such as SMASH [18], and models by Gärling et al. [20] and Vause's [21]. It is similar to these models in a sense that an activity agenda is assumed to exist, that a sequential approach is adopted to mimic the decisions involved in activity scheduling, that a "meta" decision process exists to control the flow of decisions (similar to Vause) and that alternatives to utility maximization are proposed. It goes beyond previous models, however, in terms of how priorities are assessed, how other household members schedules are incorporated in the model, how decisions are organized over time are subsequently modified during execution via the simulation model, and how the two dominant operational techniques (i.e. utility maximization versus rule-based approaches) are "unified" in the model.

Specifically, there are several aspects of the current model that make it unique from past approaches. First, it is shown how the priority of activities can be derived as a dynamic function by adopting the use of scheduling state characteristics of individuals and their household members schedules at the time of scheduling. This of particular importance for capturing the constraints imposed by other household members. Second, the current model incorporates a natural means for the rescheduling of activities that occur during execution of schedules, in the form of continuous addition, modification and deletion to the schedule. This aspect of scheduling behavior is not addressed directly in Gärling et al. and Vause's model, and is quite limited in the case of Ettema's model. This aspect of the current model largely reflects the new insights made possible by the CHASE data. Third, the current model directly addresses the sequencing of activity choices over time, something that past authors have struggled with, assuming either that decisions are made purely sequentially in time, or that some meta-decision process existed as a control mechanism. Perhaps most importantly, the current model is based on observed data that provides the necessary behavioral support of the model, and allows one to consider new types of variables for future operationalization. Both Gärling et al. and Vause stressed the importance of obtaining more data on the underlying scheduling process for future operationalization and empirical estimations of their model.

Overall, the behavioral power of the scheduling process model and simulation rests in it realistic replication of how activities are scheduled over time, how a sub-optimal solution is achieved, the allowance for a variety of decision rules, and the sensitively of the priority model over time. The priority function allows certain activities to jump up in priority depending on the circumstances. This allows infrequent, discretionary, or otherwise unusual activities to emerge depending on the situation, contributing to complex activity-travel patterns. The priority model also inherently determines the sequencing of activities in terms of the order in which decisions are made and their order in execution, without having to explicitly determine this in the model. Overall, the variables used in the priority models that reflect the state of the schedule relative to activities on the agenda are what makes this model unique, and give it the power to explain the apparent behavioral complexities of observed activity-travel patterns. The ability to collect these variables via the CHASE survey has opened up significant opportunity for future development of this model.

Providing a Unifying Framework

Looking at the derived need for travel from the perspective of an activity scheduling process allowes new insights into how past research can be brought together into a unified modeling framework. The scheduling process was shown to separate into a dual process of pre-week routine scheduling, followed by a more dynamic process of impulsive and opportunistic planning as the schedule is executed during the week. This distinction is rather convenient from a modeling perspective, as it provided a logical and behaviorally sound way to unify past econometric approaches for the modeling of routine activities, with the "rule-based" simulation approached adopted in this paper for modeling the more dynamic weekly decision process. This goes a long way towards addressing the concerns of Ettema and Timmermans [1] that a *unifying framework* which links the research in different areas is still missing.

In particular, the STARCHILD [8, 9] model appears clearly amenable to providing a model for the creation of a "skeleton" schedule of routine activities. Although in theory, the STARCHILD model makes a distinction between "planned" and "unplanned" activities, in operationalization, the model produced an activity pattern "that can be expected to be executed during the action period" [8, p. 314], and that is "sensitive to the possibility of unforeseen events arising" [9, p. 327] - very similar to the "skeleton" schedule proposed in the current model that forms the basis for unplanned scheduling during execution . The only significant modification would be to restrict the generation of feasible activity patterns to "routinized" activities, and leave "flexibility" in the form of open time periods for the remaining "unplanned" activities scheduled during the week. Tour-based models, such as that of Bowman and Ben-Akiva [2, 11, 12] could also provide the needed "routinized" framework to start the scheduling process. Although not shown, it is suspected that routinized activities are related to the "primary" tour of the day. Thus, a tour-based model could be used up to the point where the primary tour is developed via utility maximization. This would partially minimize the computational problems exhibited in these models, and provide a more solid behavioral basis for them as they are restricted to scheduling activities that do indeed lend themselves to optimization.

Future Model Development

The model proposed in this paper seeks to provide a framework for the long term development of an operational household activity scheduling model capable of outputting details on the travel demands in urban area and examining the impacts of emerging policy issues. The most immediate future research needs related to operationalization of specific components of the model using CHASE data. This includes the development of a discriminant function for routine activities and definition and estimation of the momentaneous priority function. Equally important is the identification of the types of decision rules underlying the variety of scheduling steps incorporated in the model and an assessment of how they might differ across individuals and situations. This would involve more in-depth probing using techniques such as "thinking aloud" pioneered in psychology. Additional needs include the development of a microsimulation model of activity agendas that includes the relevant attributes necessary for the discriminate and priority functions, and collaborative efforts focussing on the unification of existing optimization models for modeling the weekly routinized "skeleton" schedule. Overriding these developments is a need to develop a computer algorithm capable of simulating the scheduling process in all its components for all people in an urban area.

The ultimate future task is the integration of the scheduling modeling within a larger integrated urban model. The most obvious linkage is through the output of household level travel demands by time of day and day of week, as depicted in Figure 1. Such an effort would drastically improve the models ability to predict the impact of a wider range of policies and urban form scenarios, as well as provide inputs and feedbacks to other modeling components.

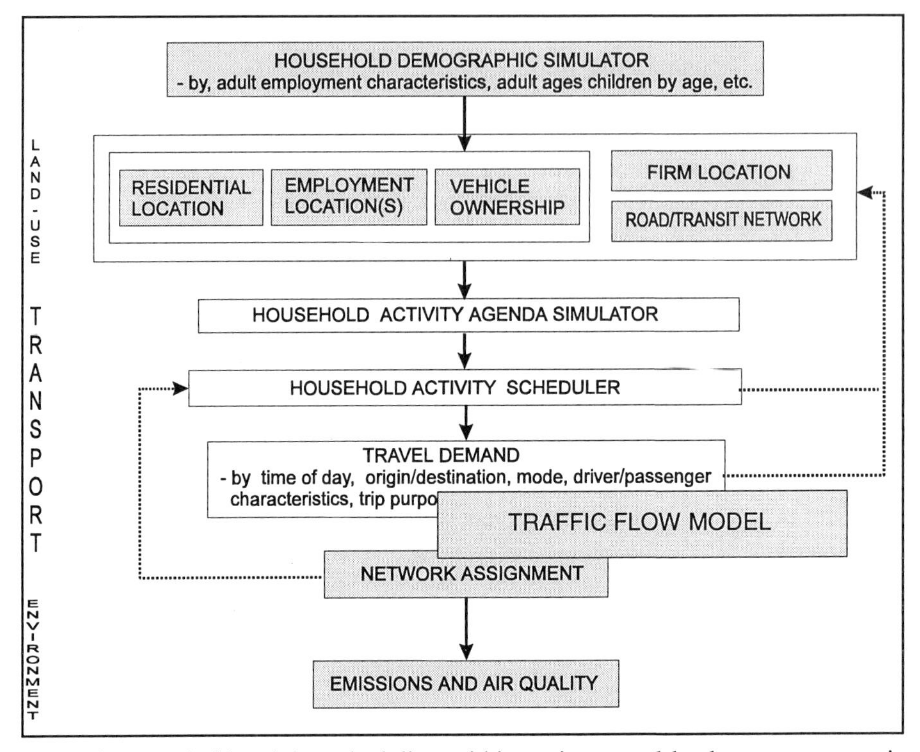

Figure 1: Household activity scheduling within an integrated land-use, transportation and environment modeling framework.

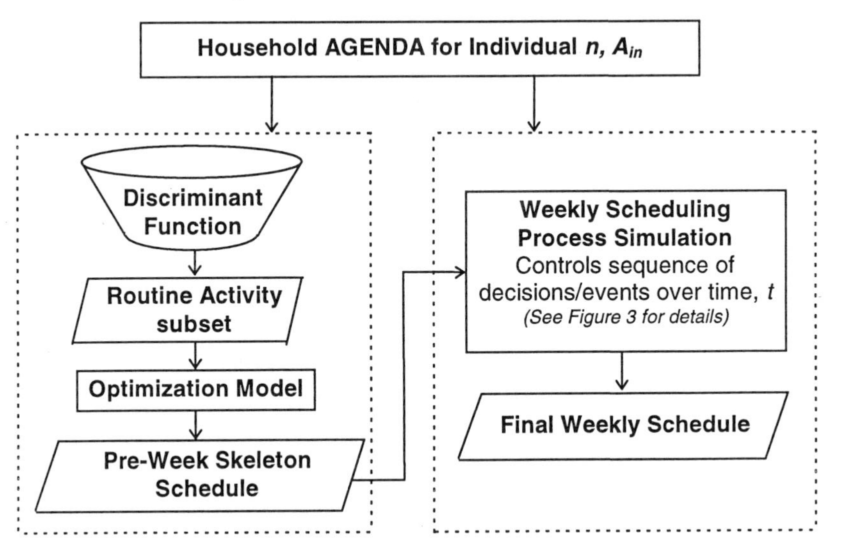

Figure 2: Weekly household activity scheduling process model, showing three major components.

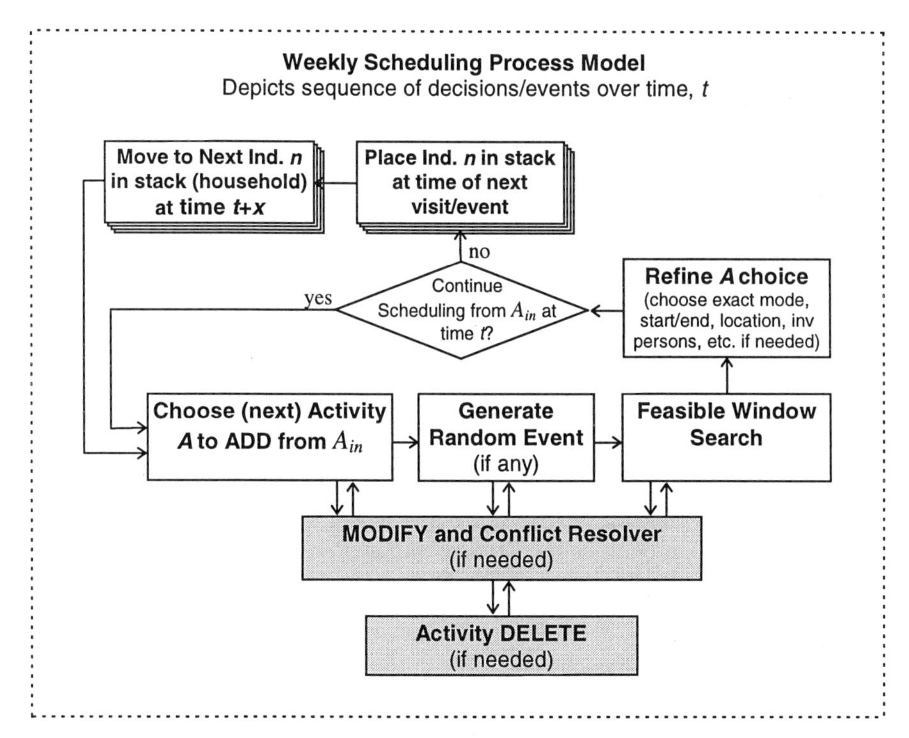

Figure 3: Weekly scheduling process model.

4 References

1. D. Ettema and H. Timmermans, Theories and models of activity patterns. In: *Activity-based approaches to travel analysis,* eds. D. Ettema and H. Timmermans, 1-36 (Pergamon, Oxford, 1997).

2. J.L. Bowman and M. Ben-Akiva, Activity-based travel forecasting. In: *Summary, Recommendations and Compendium of Papers from the Activity-based Travel Forecasting Conference*, June 2-5, 1996, 3-36. Sponsored by the Travel Model Improvement Program (1997).

3. K.W. Axhausen and T. Gärling, Activity-based approaches to travel analysis: conceptual frameworks, models, and research problems. *Transport Reviews* 12(4): 323-341 (1992).

4. R. Kitamura, C. Chen, R. Pendyala and R. Narayanam, Micro-simulations of daily activity-travel patterns for travel demand forecasting. Paper presented at *The Eighth Meeting of the International Association of Travel Behaviour Research,* Austin, Texas. September 21-25 (1997).

5. R. Kitamura and M. Kermanshah, Identifying time and history dependencies of activity choice. *Transportation Research Record* 944: 22-30 (1983).

6. T. van der Hoorn, Development of an activity model using a one-week activity-diary data base. In *Recent Advances in Travel Demand Analysis*, ed. S. Carpenter, and P. Jones, 335-349 (Gower, Aldershot, 1983).

7. P.M. Jones, M.C. Dix, M.I. Clarke and I.G. Heggie, *Understanding travel behaviour* (Gower, Aldershot, 1983).

8. W.W. Recker, M.G. McNally and G.S. Root, A model of complex travel behavior: Part I - Theoretical development. *Transportation Research* **20A**(4): 307-318 (1986).

9. W.W. Recker, M.G. McNally and G.S. Root, A model of complex travel behavior: Part II - An operational model. *Transportation Research* **20A**(4): 319-330 (1986).

10. S. Kawakami and T. Isobe, Development of a one-day travel-activity scheduling model for workers. In *Developments in dynamic and activity-based approaches to travel analysis*, ed. P. Jones, 184-205 (Avebury, Aldershot, 1990).

11. J. Bowman and M. Ben-Akiva, Activity-based model system of urban passenger travel demand. Paper presented at the *Transportation Research Board 74th Annual Meeting*, January 22-28, Washington, DC (1995).

12. J. L. Bowman, M. Bradley, Y. Shiftan, T. K. Lawton and M.E. Ben-Akiva, Demonstration of an activity based model system for Portland. Paper presented at *The 8th World Conference on Transport Research*, Antwerp, June (1998).

13. T. Gärling, Behavioural assumptions overlooked in travel choice modelling. Paper presented at the *Seventh International Conference on Travel Behaviour*, Santiago, Chile, June 13-16 (1994).

14. J.W. Payne, J.R. Bettman, E. Coupey and E.J. Johnson, A constructive process view of decision making: multiple strategies in judgment and choice. *Acta Psychologica* 80: 107-141 (1992).

15. B. Hayes-Roth and F. Hayes-Roth, A cognitive model of planning. *Cognitive Science* **3**: 275-310 (1979).

16. T. Gärling, M.-P. Kwan and R.G. Golledge,. Computational-process modelling of household activity scheduling. *Transportation Research* **28B**(5): 355-364 (1994).

17. T. Gärling, M.-P. Kwan and R. G. Golledge, Computational-process modelling of household travel decisions using a geographic information system. *Papers in Regional Science* **73**(2): 99-117 (1994).

18. D. Ettema, A. Borgers and H. Timmermans, Simulation model of activity scheduling behavior. *Transportation Research Record* 1413: 1-11 (1993).

19. C.G. Lundberg, On the structuration of multiactivity task-environments. *Environment and Planning* **20A**: 1603-1621 (1988).

20. T. Gärling, T. Kalén, J. Romanus and M. Selart, Computer simulation of household activity scheduling. *Environment and Planning A* **30**: 665-679 (1998).

21. M. Vause, A rule-based model of activity scheduling behavior. In: *Activity-based approaches to travel analysis,* eds. D. Ettema and H. Timmermans, 73-88 (Pergamon, Oxford, 1997).

22. D. Ettema, A. Borgers and H. Timmermans, Using interactive computer experiments for identifying activity scheduling heuristics. Paper presented at the *Seventh International Conference on Travel Behaviour*, Valle Nevado, Satiago, Chile, June 13-16 (1994).

23. E.I. Pas, State-of-the-art and research opportunities in travel demand: another perspective. *Transportation Research A* **19**: 460-464 (1985).

24. J. Polak and P. Jones, A tour-based model of journey scheduling under road pricing. Paper presented at the *73rd Annual Meeting of the Transportation Research Board*, Washington, D.C. (1994).

25. M. Wegener, Current and future land use models. Paper presented at the *Travel Model Improvement Program's Land Use Models Conference*, Dallas, Texas, February 19-22 (1995).

26. M. Wegener, Applied models of urban land use, transport and environment: State of the art and future developments. To appear in: *Network Infrastructure and the Urban Environment:* Recent *Advance in Land-use/Transportation Modeling* (Sprinter Verlag, New York, 1998).

27. C. Barrett, K. Berkbigler, L. Smith, V. Loose, R. Beckman, J. Davis, D. Roberts and M. Williams, *An operational description of TRANSIMS* (Los Alamos National Laboratory, New Mexico, 1995). Report LA-UR-95-2393.

28. L. Smith, R. Beckman, K. Baggerly, D. Anson and M. Williams, *TRANSIMS: TRansportation ANalysis and SIMulation System: Project summary and status.* (Los Alamos National Laboratory, Mexico, 1995).

29. R. Beckman, TRANSIMS methods. Presentation at the *Activity-Based Travel Forecasting conference*, June 2-5, New Orleans, Louisiana (1996).

30. K.W. Axhausen and P.B. Goodwin, EUROTOPP: Towards a dynamic and activity-based modelling framework. In *Advanced Telematics in Road Transport*, 1020-1039 (Elsevier, Amsterdam, 1991).

31. E.J. Miller and P.A. Salvini, A microsimulation approach to the integrated modeling of land use, transportation and environmental impacts. Paper presented at: *The 76th Annual Meeting of the Transportation Research Board*, Washington, D.C., January 11-15 (1998).

32. S.H. Putnam, EMPAL and DRAM Location and Land Use Models: an overview. Paper presented at the *Land Use Modeling Conference*, Dallas, TX, February 19-21 (1995).

33. S.T. Doherty and E.J. Miller, Tracing the household activity scheduling process using a one week computer-based survey. Paper presented at *The Eighth Meeting of the International Association of Travel Behaviour Research,* Austin, Texas. September 21-25 (1997).

34. G.C. Winston, *The Timing of Economic Activities.* Cambridge: Cambridge University Press (1982).

35. J. Supernak, Temporal utility profiles of activities and travel: Uncertainty and decision making, *Transportation Research B*, **26B** (1): 61-76 (1992).

36. T. Gärling, M.-P. Kwan and R. G. Golledge, Computational-process modelling of household travel decisions using a geographic information system. *Papers in Regional Science* **73**(2): 99-117 (1994b).

37. I. Cullen and V. Godson, Urban networks: the structure of activity patterns. *Progress in Planning* **4**(1): 1-96 (1975).

38. R. Kitamura, Sequential, history-dependent approach to trip-chaining behavior. *Transportation Research Record* 944: 13-22 (1983).

39. K.W. Axhausen, A simultaneous simulation of activity chains and traffic flow. In *Developments in dynamic and activity-based approaches to travel analysis*, ed. P. Jones, 206-225 (Avebury, Aldershot, 1990).

40. K.W. Axhausen, J. Polak, M. Boltze and J. Puzicha, Effectiveness of the parking guidance system in Frankfurt/Main. *Traffic Engineering and Control*, **35** (5): 304-309 (1994).

The Development of ALBATROSS: Some Key Issues

T.A. Arentze[1], F. Hofman[2], C.H. Joh[1], and H.J.P. Timmermans[1]

[1]Eindhoven University of Technology, Urban Planning Group, P.O. Box 513, Mail Station 20, 5600 MB Eindhoven, The Netherlands

[2]Ministry of Transport, Public Works and Water Management, P.O. Box 1031, 3000BA Rotterdam, The Netherlands

This paper discusses some experiences with the development of ALBATROSS, a rule-based system for predicting transport demand, currently under development for the Dutch Ministry of Transport, Public Works and Water Management. The model belongs to the class of activity-based models, implying that it attempts to predict which activities are conducted where, when, with whom, for how long, and the transport mode involved. In principle, this increased complexity allows one to predict the impact of urban and transport policies and institutional change on activity patterns and hence transport demand, but this increased complexity also involves new theoretical, and methodological challenges and problems of data collection. Some of these challenges are briefly discussed in this paper. In particular, the conceptualisation of activity behaviour, the derivation of choice heuristics from diary data, the development of appropriate goodness-of-fit measures and the problem of data quality are discussed.

1 Introduction

Activity-based models of transport demand have regained considerable popularity recently. This increased popularity reflects the opinion that activity-based models of transport demand potentially better capture the complex interrelationships between activity choice, choice of transport mode, destination choice and activity timing decisions, and hence may result in either better predictions or allow the assessment of different policies, which are difficult or impossible to evaluate by the existing trip or tour-based models. Various approaches, some theoretical, some operational, have been suggested in the literature to model activity patterns. An overview is given in [1].

To explore the potential of activity-based models, the Dutch Ministry of Transport, Public Works and Water Management has commissioned EIRASS/Urban Planning Group of the Eindhoven University of Technology to develop a model, which has been given the acronym ALBATROSS, A Learning-based Transportation Oriented Simulation System. Before commissioning the development of this system, a literature review was conducted and a series of workshops was held to assess the potential of activity-based models. Currently, transportation forecasting by the Ministry is based on a national model, which has been improved over the years, and which consists of a combination of (nested) logit and stated preference models. It is fair to say that it probably represents one of the most advanced models of its kind. Still, the way it was constructed involved some limitations. For example, task allocation within households, institutional change, and activity substitution could not be readily incorporated within the model system, and required ad hoc and hence largely untestable assumptions. It was this very reasoning that led to the decision to explore the potential of an activity-based model [2]. One alternative would have been to stay relatively close to the existing model, and further develop the existing model system from a tour-based to an activity-based nested logit model. However, given the explorative nature of the project, it was decided to examine the potential of a completely different approach; that is, a rule-based system. Since there is a virtual lack of such models in activity analysis, the development of the model system involves a series of largely new theoretical, methodological and data problems, for which solutions had to be found or alternatives had to be explored.

This paper reports about some of the experiences with developing this model system, which is still under development. In particular, the conceptualisation of activity behaviour, the derivation of choice heuristics from diary data, the development of appropriate goodness-of-fit measures and the problem of data quality are discussed.

To this effect, the paper is organised as follows. First, we shall briefly summarise the conceptualisation underlying the model and the essence of the modelling approach that we intend to apply. These sections should serve to better understand the model and appreciate the nature of our discussion. This is followed by a discussion of our experiences, focusing on some typical problems we encountered and the options we explored to find a solution to these problems. The paper will be concluded by discussing some avenues of future research.

2 Conceptual Framework

The basic assumption underlying ALBATROSS is that daily activity patterns and related transport demand reflect the outcome of a complex decision making process, involving the interplay between individual preferences and the opportunities and constraints induced by a specific spatial-temporal setting, which in turn might be influenced by a particular institutional context. Travel demand is viewed as a derivative of the process by which individuals schedule their activities within a given period of time, within particular household, institutional and spatial-temporal constraints, to satisfy particular goals. It is postulated that activity participation, allocation and implementation fundamentally take place at the level of the household. It is at that level that particular activities need to be performed, and it is also the household that is involved in the decision in which activities to participate. The actual generation and execution of activity calendars, programmes and schedules covers a multitude of time frames. First, long term decisions made at the household level strongly influence the generation and composition of activity calendars. Decisions regarding marital status, level of education, number of children and the like, are irreversible or require years to change, have a fundamental effect on the number and kinds of activities that need to be performed and also heavily influence the constraints that individuals and households face. These variables are also assumed to have an effect on the discretionary activities, reflecting assumed relationships between socio-demographic variables and lifestyle. Other long-term decisions, such as choice of residence, choice of work/workplace, and availability of transport modes can in principle be changed in the short run, but in general represent the kind of choices that are not changed immediately, and hence these decisions exert a strong influence on possible activity patterns as the location of the residence and workplace vis-à-vis the transportation system represent the main locations in any activity pattern and are the cornerstones of decisions.

These long-term decisions will influence household activity participation decisions. It is up to household members to allocate these activities to household members. The actual allocation will reflect task allocation mechanisms within the household, which will depend on gender-specific roles, but also time pressures. The task allocation and related allocation of activities define the possible set of activities that needs to be completed within a particular time horizon. It results in an individual activity program that is derived from the household activity calendar. We postulate that this process of program generation depends on the nature of the activities (mandatory versus discretionary), the urgency of completing a particular activity on a specific day as a function of the history of the activity scheduling and implementation process and the desire to meet particular activity and time-related objectives. Once the individual activity program has been generated, the next step is to schedule these activities, which involves a set of interrelated decisions regarding the choice of location where to conduct a particular activity, the transport mode involved, the choice of other persons with whom to conduct the activity, the actual scheduling of activities contained in the activity program, and the choice of travel linkages which connect the activities in time and space. These activity scheduling decisions thus transform an individual's activity program into an activity pattern, which is an ordered sequence of activities and related travel at particular locations, with particular starting times and duration, with particular transport modes and perhaps co-ordinated with the activity patterns of other individuals. In this context, travel decisions represent a sub-decision. Transport mode decisions dictate the action space within which individuals can choose locations to conduct their activities. The organisation of trips into chains allow individuals to conduct more activities within a specific time frame.

The actual process of scheduling activities is conceptualised as a process in which an individual attempts to realise particular goals, given a variety of constraints that limit the number of feasible activity patterns. First, *logical constraints*, such as not having breakfast late at night, limit the time slots within which particular activities can be performed. Secondly, *institutional constraints*, such as opening hours of shops, influence the earliest possible and latest times when particular activities can be implemented. *Household constraints*, such as bringing children to school, dictate when particular activities need to be performed and others cannot be performed. *Space con-*

straints also have an impact in the sense that either particular activities cannot be performed at particular locations, or individuals have incomplete or incorrect information about the opportunities that a particular location may offer. *Time constraints* limit the number of feasible activity patterns in the sense that activities do require some minimum duration and both the total amount of time and the amount of time for discretionary activities is limited. Finally, *spatial-temporal context constraints* are critical in the sense that the specific interaction between an individual's activity program, the individual's cognitive space, the institutional context and the transportation supply environment may imply that an individual cannot be at a particular location at the right time to conduct a particular activity. We assume that many of these constraints are not crisp in the sense that if these constraints are not met, a particular activity program is infeasible. In contrast, we assume that many of these constraints represent fuzzy concepts.

Having identified these constraints, the next question then is how individuals choose between feasible activity patterns. Unlike other models, which relied on utility-maximising theory, we postulate that activity patterns will evolve through *learning mechanisms*. Every time an individual tries and experiences a particular profile, positive experiences will reinforce the individual's behaviour, whereas negative experiences will reduce the repetitive implementation of the activity profile. We assume that this learning process will gradually translate active search patterns into preferences for particular locations, times and transport modes to conduct particular activities. These preferences guide decision making in the construction of feasible schedules. Hence, we reject the notion that individuals are involved in a systematic comparison of all possible activity patterns. Instead, we assume that through learning processes individuals develop behavioural rules at various levels of specificity to deal with variations in circumstances. On the one hand, individuals learn to filter out characteristics of decision alternatives that are not essential for a given scheduling purpose. The result of this generalisation is a set of relatively generic rules that allow flexible behaviour in dynamic environments. On the other hand, existing rules may turn out to be overly generalised in the sense that they produce inadequate responses under exceptional conditions. Such conflicts invoke the forming of more specialised rules that can account for exception conditions. The result of rule generalisation and specialisation is a rule hierarchy with at the top level highly generic rules and at the bottom level highly specialised rules.

3 ALBATROSS

The following requirements underlie the development of the ALBATROSS system: (i) it should be able to produce feasible schedules, i.e. schedules that are executable given the constraints imposed by the spatial-temporal setting; (ii) it should be able to produce satisfactory schedules, i.e. schedules that meet, as much as possible, preferences regarding alternative activity patterns (relaxable constraints); (iii) it should be reactive and robust, i.e. be able to react to events during schedule execution; (iv) it should be adaptive, i.e. be able to respond to environmental changes; (v) it should meet cognitive constraints, i.e. make use only of psychologically plausible mechanisms of problem solving and learning. To meet these requirements, we approached the problem as a constraint-directed search process. In this process, the system tries to satisfy preferences and continuously tests solutions against a set of explicitly represented constraints. In constraint-based search constraints tend to propagate across the system as scheduling proceeds, in the sense that one decision often limits the freedom of choice for subsequent decisions. For example, the selection of a start time for a particular activity constrains the availability of feasible/preferred start times for the subsequent activity. Given the existence of such interactions, it is not feasible for non-trivial problems to develop rules that can construct satisfactory schedules through a linear decision making process. Instead, the rules used in the scheduling system are heuristic in the sense that they often but not necessarily produce suitable solutions. Each decision is tentative and is maintained as long as it does not produce conflicts or opportunities for improvement. Since decisions are inherently tentative, a conflict-resolution mechanism that allows the system to revise decisions is built in. The same mechanism also allows the system to adjust schedules during execution when unplanned events create conflicts or new opportunities.

ALBATROSS consists of several components which are specialized in performing a certain scheduling task (Fig. 1). The *schedule representation* is the most central component of the system. It stores the current state of the evolving schedule in terms of selected and available options for decision making. The component is initialized for a given case with the given list of activities to be scheduled and available options for decisions on the timing, location, accompanying person and other activity dimensions. The *inference engine* controls the scheduling process. The engine determines a dynamic sequence in which decisions are to be made and makes sure that the various components act together to solve the scheduling problem. The engine simulates a sequential decision making process, for example, in terms of the steps: (i) allocate activities to household members and for each member (ii) determine the priority of each activity for adding to the schedule; (iii) determine for each activity in the order of priority the schedule position; (iv) specify the activities in terms of location, transport mode, exact timing and accompanying persons; and (v) revise the schedule during execution. Each of these major steps may in turn be further decomposed into subproblems which are easier to solve and so on. The resultant decision tree is dynamic in the sense that the system is able to backtrack and revise earlier decisions.

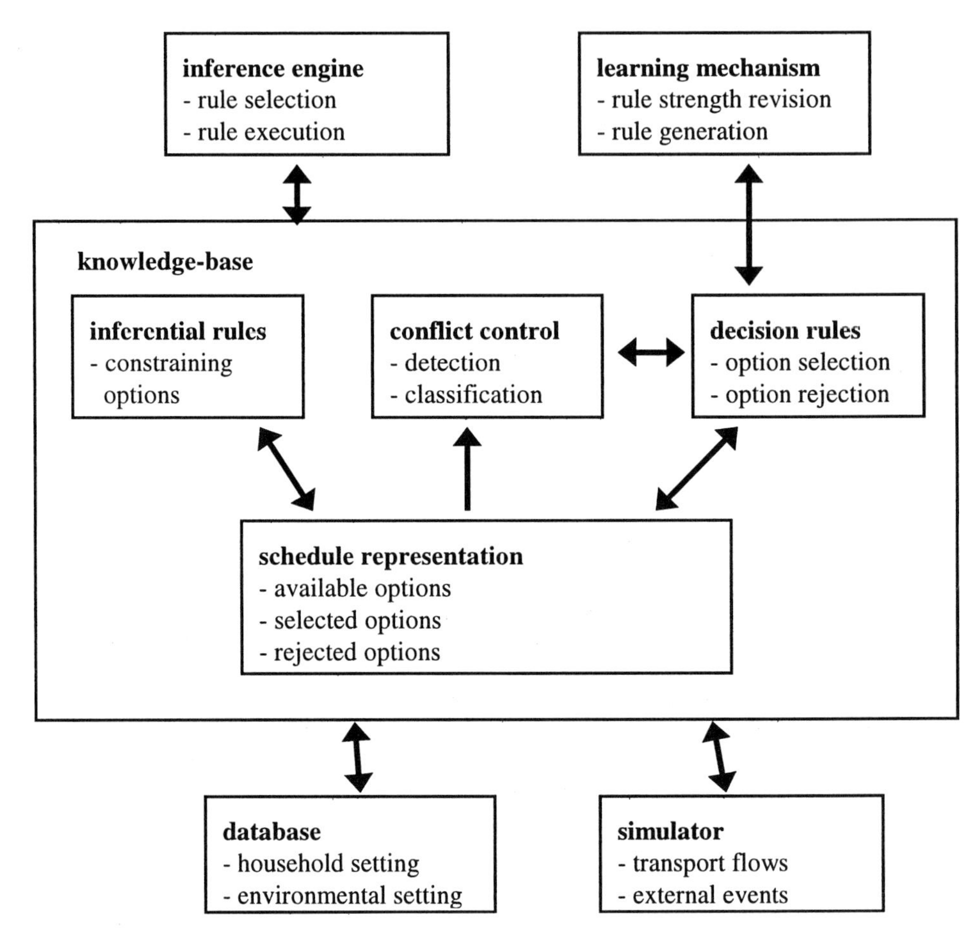

Figure 1: Structure of the activity scheduling and execution system.

Inference rules represent dynamic constraints imposed by the institutional and space-time environment. The component comes into operation each time an option is selected, to derive implications for availability of remaining options. Specifically, the rules update the availability status of options in the schedule representation component. The *conflict control mechanism* responds to possible conflicts in the schedule representation. A conflict occurs when, as an implication of a decision, a selected option on a related dimension is no longer available or proves to be suboptimal as other options become available. Having detect a conflict, this component then classifies the conflict in terms of problem categories that can be handled by decision rules.

Decision rules, which are stored in a separate component, represent decision heuristics of an individual or group of individuals. For the ability of the system to respond unambiguously under any possible condition it is important that the rules are formulated in terms of exhaustive and exclusive sets of decision rules. Therefore, the decision table is used as a formalism to represent a rule-set for each decision step. The decision tables define positive (selection) as well negative (rejection) decisions. Negative decisions are made in response to conflicts identified by the conflict control mechanism. Note that solving a conflict often involves a choice, as there are generally different ways to solve a conflict. For example, a begin time conflict of an activity might be solved by choosing a faster transport mode or reducing the duration of preceding activities. By explicitly rejecting options, the system keeps track of revised selection decisions and, thus, prevents the system from getting in an endless loop.

The *database and simulator* both represent the environment in which the scheduling system operates. The database stores information used by the knowledge-base and represents the given spatial-temporal setting. The simulator predicts dynamics of the environment in order to give feed-back on schedules generated by the system. Particularly, these predictions concern the consequences of generated activity patterns in terms of flows on the transportation network and external events. This information is fed-back to the system and usually invokes schedule revision decisions and learning processes.

Finally, the *learning mechanism* makes sure that the system is able to adapt its rule-base on feed-back on generated schedules. Adaptation takes the form of optimizing the structure of decision tables. Decision table structures can be varied systematically by merging and splitting condition states and varying the selection of actions under each (newly formed) condition state. Learning involves continuously testing small changes and reinforcing changes which generate positive feed-backs.

To summarize, the system solves scheduling problems by repeatedly passing through cycles of selecting and rejecting options (decisions), updating the availability of options (inference) and detecting conflicts (the conflict-control subsystem). These tasks are performed by specialized (but not autonomous) agents which act on a central representation of the evolving schedule. Environments are simulated to test rules and to allow the system to adapt its rule-base to changes.

4 Experiences: Conceptualisation

The way individuals and households organise their daily activities in time and space within the opportunities and constraints set by the urban environment, the transportation system and the institutional context constitutes a highly complex decision-making problem. It goes without saying that our conceptualisation necessarily has to simplify the problem. This also means that alternative conceptualisations are possible. In any case, our experiences indicate that during the process of model development we constantly needed to go back to the original conceptualisation to judge whether we should adjust it in light of particular modelling issues. In general, these decisions were related to the question at which stage of the assumed decision-making process we should incorporate particular decisions, and the question to what extent the same modelling techniques should be used for the various components of the system. To illustrate these conceptual problems, we will discuss the issue of task allocation within the household.

Our conceptual framework assumed that a household will allocate particular tasks to its members, resulting in an activity programme for each household member. While task allocation can be observed from household activity diaries, it is very difficult, if not impossible, to infer the underlying mechanisms. We therefore decided to develop an experimental approach to measure task allocation between the spouses in a household as a function of a set of conditions [3]. We decided to follow the principles of stated preference analysis, but had to change the experimental task and the estimation because the dependent variable is different from those typically used in stated preference techniques. This decision led to two fundamental questions: (i) should we develop a separate stated preference model for task allocation, or alternatively model task allocation as yet another dimension in the activity scheduling phase, and (ii) should we use a utility-maximising approach to model task allocation (as this is the theoretical premise underlying stated preference models) or, to remain consistent, develop a rule-based model.

The difficulty with questions like these is that one more or less has to develop completely different model specifications to examine to what extent such alternative conceptualisations influence the predictive ability of the model. This is not what we did, and we do not have any intention of developing complete alternatives, but testing alternative ways of developing modules is still on our research agenda. For example, rather than estimating logit models from experimental design data, rules might be derived from that data. When considering different conceptualisations of the problem, we always decided to stick with the original conceptualisation, as outlined in the previous section. A separate module for predicting task allocation has the potential advantage that the outcome of the activity sequencing decision, especially if a particular activity cannot be conducted on a particular day, can be incorporated into the task allocation model during the simulation. This seems an important property if one is interested in simulating how daily activity programmes, transport demand and scheduling decisions shift as a result of unexpected events. Whether this is true is still open for empirical testing.

5 Experiences: Deriving Choice Rules

As explained, ALBATROSS is a rule-based system. In fields other than transportation with a longer history of building rule-based systems, the rules are often elicited from experts. In our case, we wished to derive the rules from activity diary data. This created the problem of how to derive the rules. One option would be to use neural networks, but this has the possible disadvantage that the resulting network is difficult to interpret from a substantive point of view. We wished to have results that have some immediate meaning. For basically the same reason, we did rule out the use of genetic algorithms, which would try to generate strings of activities that describe observed activity patterns as closely as possible. Our research efforts concentrated therefore on the use of decision tables to represent possible choice rules and the development and test of alternative learning algorithms to derive these rules from activity diaries data. Existing knowledge-discovery and data-mining techniques for optimising decision tree structures seem to be a ready candidate for the present problem. A closer look indicates, however, that techniques such as C4.5 and AID algorithms are restricted to simple classification and selection problems. They could be used in our case if we would have observations or frequencies with which rules are selected under known conditions. Such observations are not available, as we know only outcomes of actions of rules. Moreover, the scheduling process consists of several decision steps so that the system can obtain feed-back only about the performance of collections of rules. Hence, the required algorithm must incorporate a method of apportioning credit to rules that contributed to an outcome. At the current state-of-affairs, the following approach seems promising.

The approach is based on the assumption that it is possible to specify on theoretical grounds an exhaustive set of relevant conditions as well as an exhaustive set of possible scheduling preference rules. The exhaustive sets of conditions and preferences can be represented in the form of a decision table (DT) [4]. In this 'maximum' DT, the condition section is maximally split up into columns and within each column an exhaustive set of action alternatives is available. The system incorporates for each step in the scheduling process a DT representing conditional decision heuristics for that step. By initialising each DT with the maximum DT, the entire set of possible preference rules is represented in the initial system. Then, the estimation problem reduces to optimising simultaneously for each column of each DT the selection of an action. The final step involves contracting each DT by merging conditions that do not differentiate in selected actions and removing actions that are not selected in any column of the table. A combinatory search or inductive learning algorithm can be used for simultaneous optimisation of multiple DT structures [5].

In a study conducted to test whether the proposed system is able to find patterns in diary data, we focused on the problem of determining the sequence of a given list of activities to be conducted on a particular day [6]. The diary data used represented over 2900 daily activity sequences of individuals. The problem was to find the decision heuristics that minimise an aggregate measure of distance between observed and generated schedules. For each case, the system generates a schedule by applying decision heuristics to put a given list of activities into a sequence and, next, calculates a distance measure between the generated and observed sequence. The performance of the system was measured as the aggregate distance across all cases processed. Rules were tested in a step-wise procedure. As a null model, sequences were first generated in random order. As it appeared, constraint rules were able to reduce the error of the null-model to 69.4 %. The constraint rules were a-priori designed to eliminate infeasible schedule positions for adding an activity given the time-window of that activity. The used time-windows were derived in a pre-processing step based on assumptions about the nature of activities and collected location-specific facility opening times. A further reduction was achieved by adding preference rules for the timing of eat-episodes, such as breakfast, lunch and dinner, to the rule-base. As it appears, a fixed preference time with limited tolerance for an earlier or later time gave the best results in reproducing schedules. With these added rules the base-level aggregate distance was reduced to

64.6 %. On a next level, rules representing preferred sequences with regard to the nature of activities were specified and tested. Trying to complete mandatory activities before discretionary activities and out-home activities before in-home activities turned out to be the rules that best describes the general pattern of schedules. Eventually, the most sophisticated rule-set achieved a further reduction of the base-level aggregate distance to 45.5 % and this result could be repeated on a different sample of diaries not used for estimation.

These findings suggest that even a relatively simple rule-set is able to reduce a considerable portion of error in reproducing schedules. It was not possible to further increase the performance of the system by defining groups in terms of characteristics of households, individuals or activity programs suggesting that the same rules apply to all groups. We stress, however, that the system used in this study has limited possibilities to search for an optimum. The availability of optional rules and conditions was limited by choices made by the system designer. Therefore, the next step is to extend the search space of the system and to apply an efficient algorithm to optimise DT-structures. For that purpose, another study was conducted to develop and test an inductive, incremental learning algorithm [5]. The proposed algorithm uses a probabilistic function to select rules given error scores of alternatives. The error-scores are initially set to zero and are updated each time a case is processed based on feed-back in terms of the distance measure. Simulated schedule data were used to test the algorithm and to find the optimal specification of functions used for selecting rules and updating error scores. The results of the simulation study are encouraging, but an application of the algorithm to real data has yet to be performed.

Although these preliminary results are encouraging, there are some potential limitations. One of the problems encountered is that it is difficult to define more sophisticated generally applicable rules. For example, a rule such as try to schedule a shopping activity directly after work may produce good results only in a subset of cases. In general, the problem is to find the conditions under which a rule applies. It is a matter of further research to find out whether it is possible to reduce differences in behaviour sufficiently to the conditions under which it is observed, given limited availability of data about individuals, activities and the regional setting. In particular, the possibility of defining homogenous groups of individuals with respect to preference rules might turn out to be limited. In the face of limited data availability, it is worth while considering probabilistic rules which are able to reproduce the variability of schedules that remains after accounting for known conditions.

6 Experiences: Measuring Goodness-of-Fit

In order to express the degree of correspondence between predicted and observed activity patterns, a goodness-of-fit measure is required. Given the non-statistical nature of the estimation, conventional statistical measures are not available. Hence, a descriptive measure is required. A first candidate would be the Euclidean distance or similarity measures that have found ample application in cluster analyses, but also in activity analysis. Upon reflection, however, we found those measures to be less appropriate because they do not take into consideration the sequential information that is embedded in activity patterns. Motivated by Wilson's paper in time use research [7], we decided to further explore the possibilities of the string alignment method (SAM), first introduced in biology for comparing DNA strings. SAM determines the degree of similarity between two strings of symbols in terms of the amount of effort required to make the two strings identical using insertion, deletion and substitution operations. SAM allows one to assign different weights to these operations. The weighted sum of operations required for alignment is known as Levenshtein distance. The goodness-of-fit measure is defined as the minimum Levenshtein distance across alternative paths to the solution. The equation for the Levenshtein distance is:

$$d(s,g) = \frac{1}{m+n} \cdot d(s^m, g^n)$$

$$d(s^0, g^0) = 0$$

$$d(s^0, g^j) = d(s^0, g^{j-1}) + w_i(\phi, g_j)$$

$$d(s^i, g^0) = d(s^{i-1}, g^0) + w_d(s_i, \phi)$$

$$d(s^i, g^j) = \min\,[d(s^{i-1}, g^{j-1}) + w(s_i, g_j),\; d(s^i, g^{j-1}) + w_i(\phi, g_j),\; d(s^{i-1}, g^j) + w_d(s_i, \phi)]$$

with

$$w(s_i, g_j) = \begin{cases} w_e(s_i, g_j) = 0 & \text{if } s_i = g_j \\ w_s(s_i, g_j) > 0 & \text{if } s_i \neq g_j \end{cases}$$

where: $i, j \geq 1$; $d(s,g)$ is the total cost of equalization of s ($= s^m$) with g ($= g^n$); m and n are the number of elements in sequences s and g, respectively; $d(s^i, g^j)$ is the cost of equalization of s^i with g^j, cumulated from the equalization of s^0 with g^0, w_i, w_d, w_e and w_s are the weights attached to respectively the insertion, deletion, equality/identity and substitution operations, and ϕ represents that the concerned operation is applied to no elements in the sequence.

The specific advantage of SAM for comparing sequences is that the measure is sensitive to differences in sequential information. In contrast, traditional Euclidean distance measures involve a position-based comparison of elements and are, therefore, overly sensitive to mismatches caused by one arbitrary element. Still, we had to find solutions for two specific problems that arise when one applies the sequence alignment method to activity patterns. First, although the measure takes into consideration sequential information, the deletion-insertion operation implies that the measure is insensitive to the complete structure of the sequence. The position where these operations take place do not matter. We therefore elaborated the basic method by developing a position-sensitive sequence alignment method. For example, assume we have two sequences *E*ABCD and ABC*E*D to compare with another sequence ABCD*E*. The conventional SAM gives the same alignment cost of 2 units for both comparisons. Focusing on the positions of the element *E* before and after the alignment, the conventional SAM regards the operations applied to the element *E* as a deletion and an insertion. Because those two operations are applied independently, the relative position of insertion to the position of deletion is of no concern. In reality, however, it may be argued that those two operations are connected because the insertion operation is applied to the same element that has been deleted before. The deletion of an element and the insertion of the same element therefore are not two independent actions but just one action, the 'reordering' of an element. A position-sensitive SAM may return the comparison results of the above example such that the first comparison costs 4 units for the reordering of *E*, while the second costs only 1 unit. Numerical simulations indicated that although this measure improved the sensitiveness of the goodness-of-fit measure, it did not dramatically change the results.

Perhaps more important is that SAM is a unidimensional measure, whereas of course we need to compare the predictions against the full activity profile. We therefore developed a multidimensional extension of the unidimensional SAM [8]. It provides minimum alignment costs of *K*-dimensional activity pattern pairs by integrating *K* sets of unidimensional deletion, insertion and substitution operations. The unidimensional operation sets are equivalent to the operation paths in the Levenshtein comparison table of the unidimensional SAM.

For the basic case this turned out to be a more or less straightforward solution. However, the solution caused the problem that the identification of the optimal path requires a substantial amount of computing time, which should be avoided if the model is to be used in real-time. The problem is that many combinations of paths (trajectories) are equivalent. We therefore have tried a more efficient algorithm, maintaining the method of the segments-based alignments, while ignoring the orders of the deletion and insertion operations and producing unique (non-repetitive) unidimensional operation sets. This result is achieved by tracing identity-operation-only trajectories rather than all-operation trajectories, and next, converting the identity trajectories into deletion and insertion trajectories to compute alignment costs.

Nevertheless, the number of unique unidimensional operation sets itself still remains fairly large, and many results of the integration of different unidimensional operation sets involve the same multidimensional alignment costs. Typically, series of successive elements of secondary attributes (location, transport mode, accompanying person, etc.) produce many different identity trajectories, because the specific elements of these attributes are not chosen for their own sake, but are driven by activity performance (primary attribute). A more fundamental solution is required to ultimately eliminate redundant unidimensional operation trajectories that lead to an enormous number of repetitions of the same multidimensional operation sets. We therefore are now exploring the possibility of developing a heuristic approach, using genetic algorithms (GA).

The purpose of this GA-based multidimensional SAM being developed is the same as before. That is, (1) identifying unidimensional identity trajectories for each attribute, (2) combining k unidimensional identity trajectories, one for each attribute, (3) converting each of the k unidimensional identity trajectories of the trajectory combination to a unidimensional deletion-insertion operation set, (4) creating a multidimensional operation set for each trajectory combination by integrating k deletion-insertion operation sets, and (5) computing the costs of multidimensional sets of deletion and insertion operations and choosing the minimum as the distance between two patterns.

The GA-based multidimensional SAM, however, reduces the number of trajectory combinations to test, while attempting to finally obtain an acceptable solution by probabilistically selecting solution candidates based on gradually improving fitness values. A set of genetic operators drives the above procedures, (2), (3), (4) and (5), until a predefined termination condition is met.

The GA-based SAM being developed includes three basic genetic operators. The *reproduction* operator reproduces trajectory combinations selected on the basis of fitness values derived from distances computed for the trajectory combinations. The *crossover* operator exchanges (parts of) unidimensional trajectories of an attribute between trajectory combinations selected on the basis of fitness values. The *mutation* operator replaces poor trajectory combinations with newly generated ones. All selections of trajectory combinations for reproduction, crossover and mutation are made probabilistically. The preliminary results indicate that the model still needs to be improved, especially with respect to the crossover operator.

7 Experiences: Data Quality

Activity-based models predict which activities will be conducted where, when, for how long, the transport mode involved, and sometimes with whom and the route that is involved. Consequently, data on all these choice dimensions are required. Activity diaries typically record this information and hence seem the logical choice to collect the required data. It has been recognized in the literature that this type of diary offers some potential advantages over conventional transport surveys and travel diaries (see [9] for an overview). However, while activity diaries potentially result in a more detailed recording of activities and related travel, this comes at a clear cost. The completion of activity diaries is very demanding, resulting in higher non-response rates, differential non-response by socio-economic groups and higher item non-response rates and inconsistencies. Perhaps even more importantly, activity diaries are also more prone to errors compared to conventional travel surveys. The diaries collected for developing ALBATROSS were no exception. The diaries contained considerable missing information and inconsistencies. For example, if a particular activity is conducted at the same location, respondents tend to provide the detailed information only once, assuming that the researcher can figure out what happened during the other occurrences of that activity. In any case, the quality of the original data was such that a true effort was required to clean the data and arrive at an acceptable data quality.

Fortunately, by their very nature, activity diaries should comply with several logical constraints and relations. For example, if two consecutive activities are not conducted at the same location, one knows that travel between these two activities should be involved. Hence, we decided that the consistency of the activity diaries can in principle be verified by using a set of logical rules. In addition, missing elements of observed activity diaries can be inferred by relating different pieces of information and applying one or more decision rules to generate a logically consistent and complete activity diary.

To this effect, we developed an interactive computer system, called SYLVIA, to test the logical consistency and completeness of activity diaries and infer logically consistent and complete activity diaries [10]. This system was written in C++ under Windows95. The core of the system is represented by the DIAGNOSIS/REPAIR function. Both apply the set of rules underlying the system to the data. In the diagnosis mode, the system will identify and count those respondents or data entries who fail to satisfy particular rules. In the repair mode, activity schedules will be corrected according to the rule of interest. In most cases, the user is not given an option, but alternatively a choice can be made between alternative correction rules. The results of the diagnosis and repair routine are displayed in terms of the frequencies of a match for each rule. Optionally, cross-tabulations can be generated such as for example a breakdown of errors by some classification of households, allowing the analyst a better understanding of the data quality for various segments of the sample. Another module is the ANALYSIS option. It can be used to analyze the relationship between diary and region data for example in terms of activity location versus available facilities, begin and end times versus opening hours of facilities, reported versus objective travel times and so on. Using this option one can identify and locate systematic discrepancies between region and diary data. Our experiences with applying this system were positive. Many inconsistencies, especially the more important ones, are identified and corrected, and a considerable amount of missing information can be generated by using available information from other parts of the diary or from statistical information. However, our experiences also suggest that some diaries contain too much missing information or too many inconsistencies that the application of these rules, however advanced they may be, still does not completely solve the problem. In that case, one should either delete the diary or solve the problem by visual inspection.

8 Conclusions

This paper has briefly reported some experiences with developing a rule-based system of transport demand, based on activity diaries. The problems we encountered range from conceptual problems, via problems of data quality to computational problems. The present paper only comments on a small portion of these problems.

In general, the amount and nature of the problems seem typical for explorative research in which one tries a new research direction, void of considerable previous research endeavours. It implies that one has to deal with a whole series of new problems, rather than with incrementally improving or extending existing methodology. Often, there is no time to fully compare alternatives, and the conclusions therefore necessarily remain tentative.

Our experiences to date suggest that the development of an operational rule-based model of activity behaviour and related transport demand seems feasible, but that such a model is considerably more complex than discrete choice models of activity patterns. To some extent, this might be just a matter of time. Discrete choice models have been developed over the last 25 years and hence the software to estimate those models is readily available. In our case, all the software had to be created from scratch. However, the implied complexity and level of detail will always imply that researchers should spend more effort in specifying the rules, identifying the constraints and collecting the data required to apply the model.

It is to be expected that the implied complexity will allow the use of the model to assess policies for which existing models are less appropriate. Whether the increased effort is worthwhile will be open to empirical testing. We are planning to compare the performance of ALBATROSS with other activity-based models, based on nested logit specifications, in the near future.

References

1. D.F. Ettema and H.J.P. Timmermans. Theories and Models of Activity Patterns. In *Activity-Based Approaches to Travel Analysis*, eds. D.F. Ettema and H.J.P. Timmermans, 1-36. (Elsevier Science, Oxford, 1997).

2. A.W.J. Borgers, F. Hofman, and H.J.P. Timmermans. Activity-Based Modelling: Prospects. In *Activity-Based Approaches to Travel Analysis*, eds. D.F. Ettema and H.J.P. Timmermans, 339-353. (Elsevier Science, Oxford, 1997).

3. A.W.J. Borgers, F. Hofman, M. Ponjé and H.J.P. Timmermans. Towards A Conjoint-Based, Context-Dependent Model of Task Allocation in Activity Settings: Some Numerical Experiments. Paper presented at the TRB Conference, Washington, 1998.

4. G.L. Lucardie. *Functional Object-Types as a Foundation of Complex Knowledge-Based Systems*, Ph.D.-Dissertation, Eindhoven University of Technology, Eindhoven, The Netherlands, 1994.

5. T.A. Arentze, F. Hofman and H.J.P. Timmermans. Estimating a Rule-Based System of Activity Scheduling: A Learning Algorithm and Results of Computer Experiments. Paper presented at the Informs San Diego Conference, San Diego, US, 1997.

6. T.A. Arentze, F. Hofman, and H.J.P. Timmermans. Deriving Rules from Activity Diaries. Paper presented at the WTRC Conference, Antwerp, Belgium, 1998.

7. C. Wilson. Activity Pattern Analysis Using Sequence Alignment Methods. Paper presented at the Conference of International Association of Time Use Research, 1996.

8. C.H. Joh, T.A. Arentze, F. Hofman and H. Timmermans. Activity Pattern Similarity: Towards a Multidimensional Sequence Alignment. Paper presented at the IATBR Meetings at Austin, Texas, 1997.

9. T.A. Arentze, F. Hofman, N. Kalfs and H.J.P. Timmermans. Data Needs, Data Collection And Data Quality Requirements Of Activity-Based Transport Demand Models, to be published by the Transportation Research Board, 1998.

10. T.A. Arentze, F. Hofman, N. Kalfs and H.J.P. Timmermans. SYLVIA: A System for the Logical Verification and Inference of Activity Diaries. Paper presented at the TRB Conference, Washington, 1999.

Analysis of Traffic Flow of Goods on Motorways by Means of Video Data – Chances and Limits

K. Henning, K. Grobel, and H. Saß

Lehrstuhl Informatik im Maschinenbau (IMA/HDZ), RWTH Aachen, Dennewartstr. 27, 52068 Aachen, Germany

In a field study in the area of Aachen-Düsseldorf-Köln a procedure of documentation and analysis of goods traffic streams on motorways with videodata was applied. After an extensive preparation all data had been transferred into origin destination matrices. Due to this reliable statements on the traffic situation during the observation period could be made. In this way, the basic applicability of the procedure was shown

1 Introduction

The road transport is assumed to be a subsystem of traffic. Generally all model ideas of **m**otorized **i**ndividual [motor car] **t**raffic (M.I.T.) do not apply to road transport in the same way. Up to now M.I.T. can be predicted by using generation models of traffic (e.g. the 4-stage algorithm) based on data of population and economy structure [1]. These procedures have to be adapted correspondingly to forecast road transport. In this case, problems occur. A freight is often reloaded several times on the way from the consignor to the receiver and covers distances which can not be evaluated with identical criteria like M.I.T.. For instance, spatially or within a given time shortest routes are chosen in the M.I.T., if possible. Presently, the haulage is only a very small part of the cycle costs of a product, therefore long transport distances are often approved. Though this does not happen accidently but as a result of the logistics and disposition of the enterprises involved. These connections can hardly be modeled. Therefore, the procedure of forecasting of freight traffic deviates from M.I.T. [2].

The human being is in the final analysis always the essential element in the transportation system. Through him, the system is not resp. only relatively reproducible or computable. An exact forecast of states of the transportation system is not possible. This is a quality of living systems in which people as elements are contained. The great number of different elements (people and vehicles) and of their time dependened relationships characterize the transportation system and therefore also road transport as a complex system [3].

Emphasis of this work is the consideration of freight traffic data. In recent years the freight traffic on the street increased considerably, as total traffic did too. However, compared to M.I.T. the data situation of freight traffic is incomplete. One reason is that standardized recording of the truck-frontier-crossings have stopped since the aperture of the boundaries between the European Community countries.

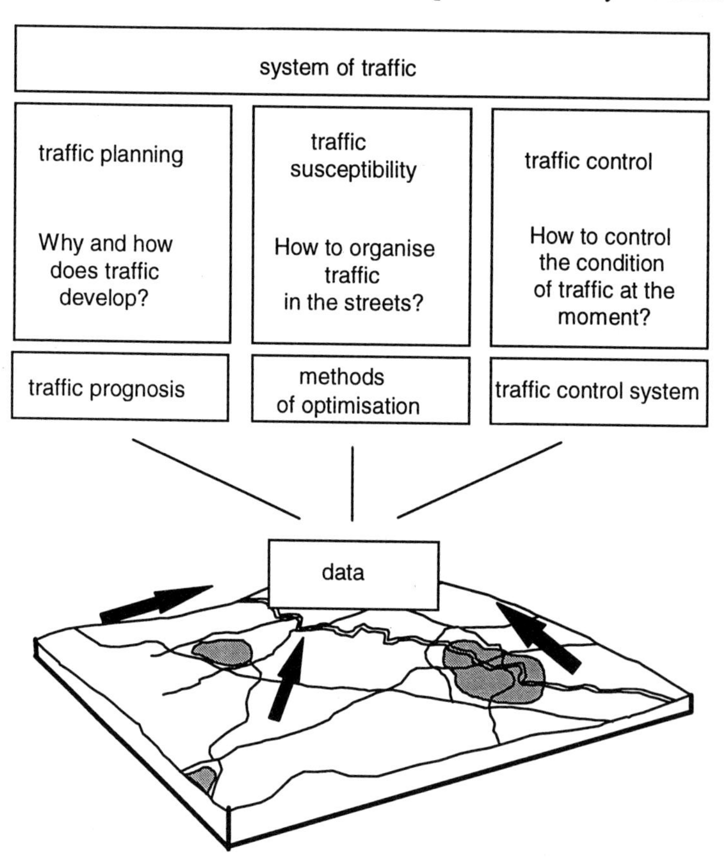

Figure 1: The investigation fields of the street transportation system.

Methods to analyze in case to influence goods traffic, e.g. simulations or traffic management systems, require a data base founded on real traffic. For this purpose (freight traffic), there are evaluations available which are usually not suitable to generate the situation for a due-date with the desired solution. For recording traffic, there is data source by continuous traffic counting points which state little about alternative route selection of the trucks.

Within traffic, the three scientific fields traffic planning, traffic susceptibility and traffic control can be distinguished. The rather long-term traffic planning is traditionally the task of the road construction engineers and town planning engineers. The development of concepts for traffic control and traffic control system is nowadays supported by electrical engineering and the control technology. Traffic affect systems attempt to combine planning approaches oriented on a long-term basis with the control technology for short-run traffic statuses. Data has to be provided for each of these tasks. Models on traffic are the tools which prepare these data for a specific application. Fig. 1 shows this division and the functions showing from it [4].

To answer these questions, suitable models on traffic have to be used as simplified "image" of the real transportation system. With this aid, the transportation system can be described. Immanent complexity could be reduced in different ways. This is achieved either by a corresponding definition of the system limits or by simplification of the dynamic connections because of the complexity. An extensive space containing for example the entire motorway network of the Federal Republic of Germany, will be much harder to describe as a motorway fragment. Fig. 2 shows three categories of the traffic consideration after the extension of the spatial system boundaries.

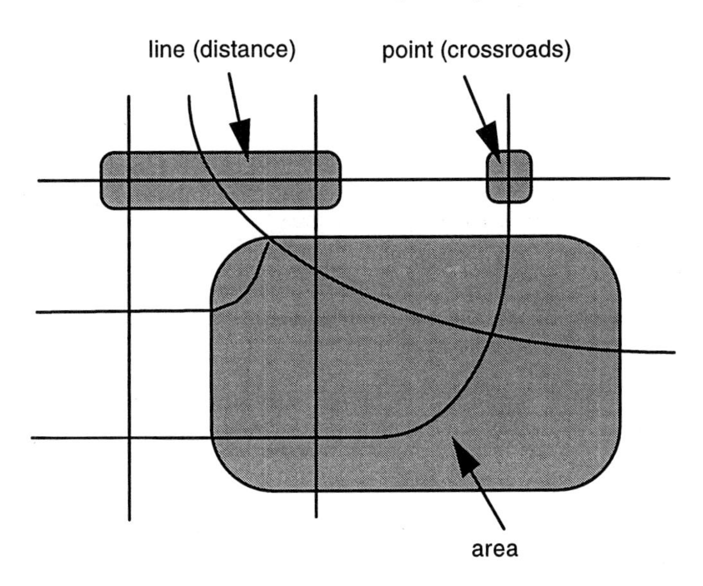

Figure 2: Possible system boundaries of an investigation field.

A transportation system can be considered in its spatial extension, which demands an exact description of the network. By reducing the system limits to a route or a crossroads the model is simplified in one dimension. *High-powered* connections, such as movement of individual vehicles, can also be simplified. In such a way for the consideration of the traffic flow, the different current rates of the single vehicles can be replaced through the middle speed rate of the vehicle-collective.

In the following two kinds of models are explained. Exogenous models make use of origin destination matrices as a tool to describe the traffic connections. The goal of endogenous model is to determine the origin destination matrix by observing real traffic densities (traffic counts). Fig. 3 shows the two counterrotating models of the

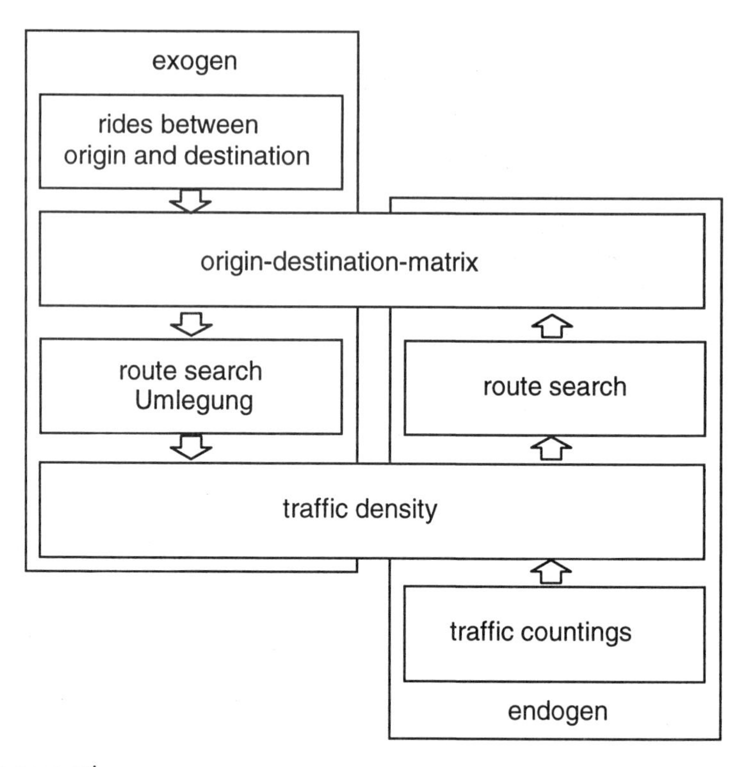

matrix generation:

Figure 3: Exogenous and endogenous models of the matrix generation (according to [5]).

Exogenous and endogenous methods can be combined fundamentally. A matrix developed exogenously can be projected or posted with the matrix estimated from counting data. In this case however, one has to accept that information gained during the data transcription (for example information about ride routes) gets lost and will not be represented through the matrix.

2 The Procedure for the Analysis of the Goods Traffic Streams

Due to the poor data situation of freight traffic and the aspiration to make most precise statements about start target relationships with a high temporal solution (a maximum of 5 minutes) the videodata consumption was selected as suitable tool. The procedure applied consists of different steps: preparation and planning; carrying out to pick up data; preparation of the data and evaluation of the data. For the preparation and evaluation of the data, a database program was developed. This program also permits the comparison of continuous counting data with videodata at further configuration levels. The functional construction of the program is represented in Fig. 4.

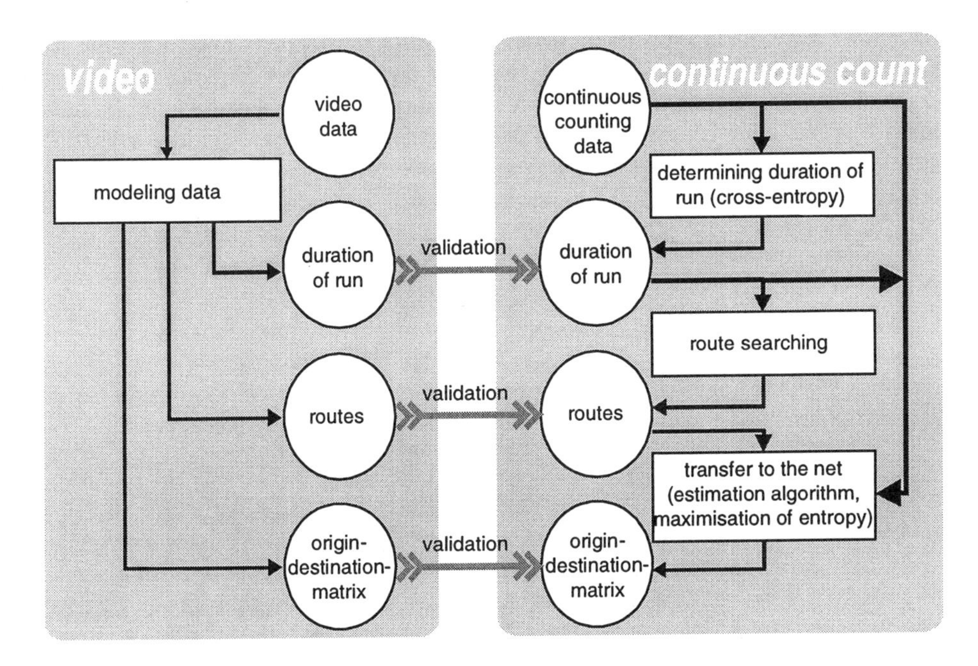

Figure 4: Construction of the program for evaluation of the videodata.

3 Planning and Carrying out of the Data Recording

The Lehrstuhl Informatik im Maschinenbau (IMA) at Technical University (RWTH) Aachen, the Institut für Kraftfahrwesen Aachen (ika, RWTH Aachen) and the Lehrstuhl für Verkehrswesen (LfV, Ruhr University Bochum) carried out a video recording of truck traffic on Monday 10 June 1996. In the following, an overall view of the planning and carrying out of the video recording should be given. This was the selection of the motorway net, the selection of the recording sites, the carrying out of test shots and finally picking up videodata and the evaluation.

During the field study a section of the motorway network in NRW was selected to catch as much data as possible. To register the traffic between the large towns too, the city triangle Aachen-Köln-Düsseldorf was chosen to be the investigation field. Furthermore, one measuring point had to be on each motorway in the investigation field, if possible near an automatic continuous counting point. However, this was only possible in a few cases, especially when a bridge over the motorway was near a continuous counting point.

In addition, the recording sites had to be placed in such a way that a truck could possible be registered several times on its way through the investigation field. By regarding available financial funds and staff resources this led to the decision to arrange 12 video measuring points.

First test shots could show the basic aptitude of standard video cameras. Further test shots and first evaluations led to a more precise formulation of demands on the camera position and the camera attitudes. VHS quality proved to be sufficient. A shutter had to be available because freezing frames where presupposed for the evaluation. Also a stable tripod was necessary to set up the camera behind the parapet of the bridge.

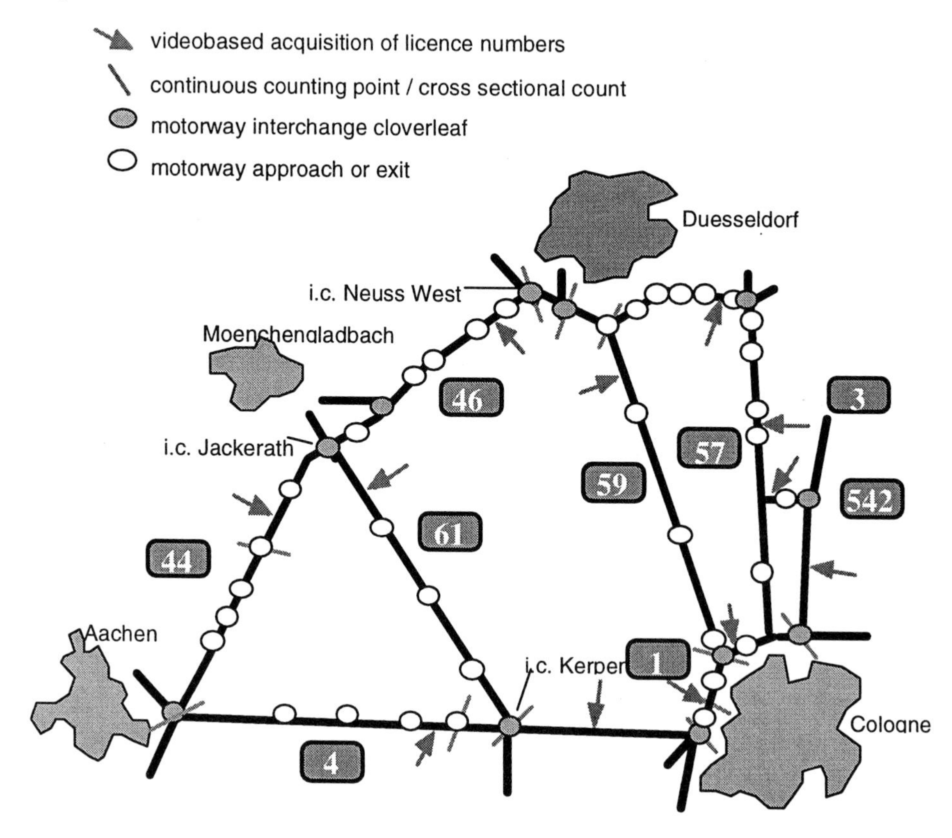

Figure 5: Photo sites of the video recording.

The first test shots had shown already that a video recording would only be able to be carried out by picking up traffic from the back. Otherwise, the cameras would have been recognized from a distance as radar control. The cameras were set up in a way that the back of the vehicles could be registrated. The trimming should only register the width of one traffic lane. This could guarantee the trucks' licence numbers would appear sufficiently large in the figure afterwards. A possibly available autofocussing had to be interrupted.

The selected sites were documented as well as the approach roads (photos, rough drafts). The following Fig. 5 gives a view of the sites of the video cameras.

For a four-lane motorway (2 lanes per direction) two video cameras were necessary to pick-up the right lane for the time being. Totally 30 cameras were in use. Every camera was mounted onto a tripod and set up in a way not to attract attention. Current supply for the four hours lasting take was guaranteed by the use of car batteries.

Altogether 40 employees, 36 of them provided by the IMA, took part in the video recording. These people were prepared for their task in a training course on 7 June 1996. The education consisted of a general part where the intention was explained and general references were given; furthermore, in a second part every group (each of them 3 persons) got opportunity to become familiar with the material (the cameras and the current supply). A manual containing most important references and information presented and passed to everybody on „V-day".

4 Preparation and Workmanship of the Data

110 hours of acceptable video material resulted from the data investigation. On account of the ADTD data (average daily traffic density in vehicles per 24 hours) of truck traffic in the investigation field coming from the year in 1991 [6], there were during a time of 6:30 to 10:30 at least 30,000 trucks expected in the investigation area.

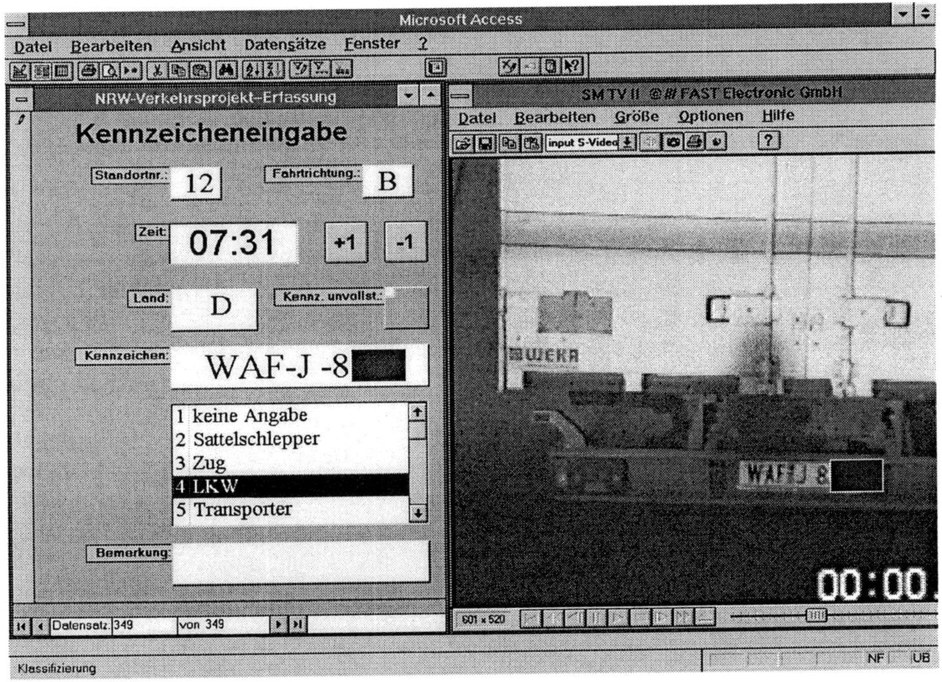

Figure 6: Input mask of video evaluation.

Possibilities offered by automatic video evaluation systems had been tested before. No usable programs which satisfied the high demands on a licence numbers recognition in an open system could be found. Therefore, the evaluation of the video recordings occurred manually through the input of the licence numbers into a database. On the basis of the selected database program (Microsoft Access 2.0), an input mask was developed. With this aid, the input of the licence numbers into the database was simplified. Presenting the video figure and the input mask on the same screen was the most favorable option (Fig. 6). The input mask allows a plausibility check on numbers from Germany, the Netherlands and Belgium. Depending on the users penchants the input mask can be operated either with the keyboard or the mouse. The time to evaluate the video tapes took, depending on the number of trucks and the practice of the person evaluating, between twice and five times longer than the recording time.

The evaluation of the data showed a number of 34,130 registered trucks so that the number of about 30,000 trucks from the ADTD values of year 1991 were exceeded according to expectations. For each driving direction of all measuring points, a table which registers the licence numbers of the trucks in minute ranges was prepared. Approximately 30,500 signs out of these 34,130 trucks could be registered in a quality that a route pursuit was possible. This corresponds to a part of approx. 89.4%.

5 Evaluation of the Data

At first the prepared data were checked regarding to their plausibility. The total number of counted trucks in direct comparison with videodata and automatic continuous counting points is to be taken from Tab. 1.

Measuring point	Videodata	Continuous counting	Difference
Kerpen	2.109	2.616	-19,4 %
Neuss-Holzheim, A	1.380	1.572	-12,2 %
Neuss-Holzheim, B	1.500	1.371	-8,6 %
Jülich-Mersch, A	1.163	1.044	+11,4 %
Jülich-Mersch, B	1.024	1.254	-18,3 %

Table 1: Comparison of videodata with the continuous counting point data.

According to expectations the number of recognized vehicles on most of the counting points is higher than in the case of videodata. There are two reasons. On the one hand videodata were taken only from the right lanes and in case of six-lane motorways for the middle lanes too. Therefore a small number of vehicles could not be registered. On the other hand a by far greater part of those vehicles which were recognized by the continuous counting points were not considered during evaluation of the videodata. This is in particular means an amount of vans with an estimated total weight up to 3.5 tons. The reversion of this ratio at the measuring point "Jülich-Mersch A" indicates probably a malfunction of the continuous counting points.

The application of the database program to generate the start target matrices determined 4,793 routes from a 30,500 recognized licence numbers. These routes are represented in the following Tab. 2.

From/to	V01	V02	V03	V04	V05	V06	V07	V08	V09	V10	V11	V12	sum
V01	0	40	31	34	2	0	10	0	10	6	17	8	158
V02	20	0	5	4	2	0	10	0	7	11	159	28	246
V03	171	55	0	6	0	0	2	0	8	3	11	16	272
V04	46	10	109	0	53	21	32	0	99	33	8	9	420
V05	3	2	10	77	0	82	22	11	7	65	5	10	294
V06	0	1	11	20	99	0	21	7	3	15	1	1	179
V07	8	5	3	57	77	32	0	41	31	173	11	15	453
V08	6	10	2	2	2	0	1	0	2	339	27	91	482
V09	19	12	69	46	20	2	19	4	0	400	4	11	606
V10	2	15	3	9	85	29	210	22	30	0	47	76	528
V11	10	62	4	9	9	5	19	6	12	41	0	192	369
V12	26	27	0	2	12	2	32	6	8	151	520	0	786
sum	311	239	247	266	361	173	378	97	217	1237	810	457	4793

Table 2: Origin destination matrix of freight traffic from videodata.

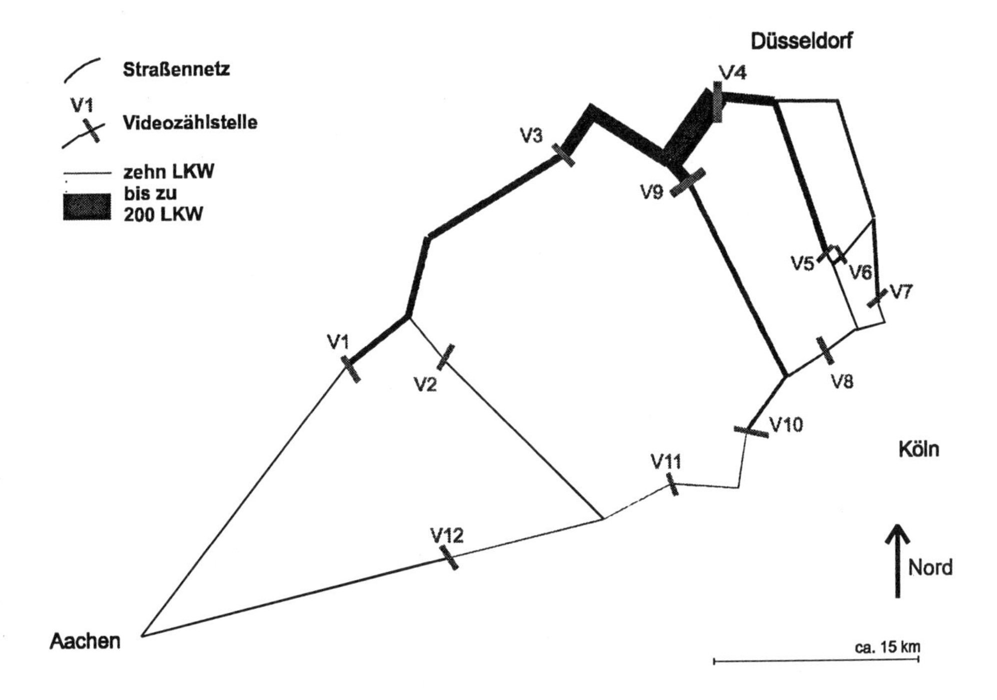

Figure 7: Distribution of source traffic of a video measuring point.

Beyond these general interlacing relationships, the routes to specific destinations of a destination can also be represented (Fig. 7). In this case it has to be noted that actual start and final destinations are of course not the video counting points; this is the same with the origin destination matrix, too. Therefore, this statement applies only to those motorway sections which combine the counting points with each others.

6 Evaluation of the Procedure

While setting up the cameras it was aspired to hide the equipment as far as possible behind the balustrades of bridges. Nevertheless, in some places traffic was affected. In places with dense traffic already (rush hour on motorway A4 near Düren and on the motorway crossing Köln) traffic blocks were specified. A slight rear-end collision was also observed. However, recording the trucks remained without effects. The decision to include only the traffic on the right lane proved as sufficient, according to expectations the number of trucks on the left lane was small. It could even be observed that truck drivers, presumably notified via citizen's band, went into the right lane when facing a video measuring point and now avoided to overtake.

The accuracy of the video pick-up procedure proved with a quota of approx. 89% for recognized licence numbers it is also possible to get a temporal solution at minute ranges. In this way, models which require precise data can apply correspondingly this method. Also the aggregated data can be prepared in different ways and get usable for further procedures.

However, the carrying out and evaluation have been extremely personnel-intensive and therefore, it is not to be recommended with present techniques. But evolution in the field of video evaluation passes that fast that one can expect systems suitable for the effort of conditions on motorways in near future.

7 Summary

Because of the video recordings during entry and exit of trucks to or off the investigation area a description of individual routes for several vehicles is now prepared. At present it is not possible to register the entire traffic with justifiable expenditure. The video pick-up supplies only a small section of traffic, however, a very precise one. Comprehensive evaluations like those introduced by the DDG at this time, are on principle conceivable if the techniques for analyzing will be improved. Moreover, statements about origin-destination relationships could be applied to traffic control and also traffic planning.

References

1. Mensebach, W.: Straßenverkehrstechnik. Werner-Verlag, Düsseldorf, 1974.

2. Schmidt, H.-G.: Erhebung und Hochrechnungsmethodik der Straßenverkehrszählung 1975 in der Bundesrepublik Deutschland. Forschung Straßenbau und Straßenverkehrstechnik, Heft 4, Bonn, 1976.

3. Henning, K., Marks, S.: Kommunikations- und Organisationsentwicklung. Vorlesungsmanuskript, Verlag der Augustinus Buchhandlung, 4. Aufl., Aachen, 1995.

4. Kreuser, J.: Darstellung und Einordnung von Verfahren zur Beschreibung der Vorgänge im komplexen System Verkehr. Studienarbeit 496 am IMA/HDZ der RWTH Aachen, 1998.

5. Keller, H., et.al.: Systemdynamische Schätzung der Matrix der Verkehrsbeziehungen in Außerortsstraßennetzen als Grundlage für die Steuerung von Verkehrsleitsystemen. Forschung Straßenbau und Straßenverkehrstechnik, Heft 702, Bonn, 1995.

6. Ministerium für Stadtentwicklung und Verkehr des Landes Nordrhein-Westfalen (Hrsg.): Straßenverkehr auf Außerortsstraßen in Nordrhein-Westfalen, Kurzbericht zur Analyse, Düsseldorf, 1992.

Estimating Path Flows from Traffic Counts

M.G.H. Bell and S. Grosso

Transport Operations Research Group, Dept. of Civil Engineering, University of Newcastle, Claremont Tower, NE1 7RU, Newcastle upon Tyne, UK

The theory and use of the Path Flow Estimator (PFE) as a one-stage network observer in support of urban traffic management and control is presented. Following an outline of the theory, initial results from the CLEOPATRA and COSMOS projects are reported. Underlying the PFE is a flexible capacitated stochastic user equilibrium traffic assignment method which provides unique estimates of path flows and path travel times from traffic count and prior origin-destination data. To accommodate transitory overloading, time-dependent delay functions are used whereby final queues in one time slice are passed forward as initial queues for the next. The PFE is programmed in C, and can run on a variety of platforms. Efficient use of memory means that large networks can be handled in real time on a PC, although the algorithm stores and processes individual paths.

1 Introduction

An important function of Intelligent Transport Systems (ITS) is the on-line monitoring of network conditions. The Path Flow Estimator (PFE) is a *network observer* that estimates network properties, in particular path flows and path travel times, from sensor data. Prior to the PFE, network observers have used a two-stage approach; first a trip table is estimated from link flow measurements assuming constant link choice proportions (namely the proportions of traffic between each origin and destination that chooses each link), then the estimated trip table is assigned to the network assuming flow-dependent link choice proportions, allowing for the effect of congestion, thereby arriving at estimates of path flows and travel times. There is a risk of inconsistency arising from the differing assignment assumptions of the two stages, unless these are reconciled by some iterative scheme. Furthermore, there are difficulties associated with the widely-used equilibrium assignment principle, associated with the lack of uniqueness of the resulting link choice proportions and the unrealistic assumption that users are perfectly informed.

The PFE, originally developed for the DRIVE 2 project MARGOT [1], is a one-stage network observer that allows path flows to be estimated uniquely and avoids inconsistency in the assignment assumptions. Trip table estimates, if required, may be obtained by aggregating path flows. It is currently being tested in a number of projects, some funded by the European 4th Framework Programme and others by the Engineering and Physical Science Research Council. Within the CLEOPATRA project, the PFE is being tested in conjunction with the urban traffic management and control systems of Lyon, Toulouse and Turin. In the COSMOS project, the PFE is being integrated in a novel traffic signal control system, referred to as MOTION, with a field trial in Piraeus, Greece. In MOTION, the PFE will provide short to medium term path flow and travel time predictions for urban traffic signal control and collective route guidance.

The core of the PFE is a logit path choice model, whereby traffic is assigned to paths on the basis of their travel times, taking congestion into account. For those links where flow measurements are available, shadow prices are calculated so as to reproduce the measurements, plus or minus a tolerance. In addition to the flow measurements, the dynamic inputs include the signal plans and one or more trip tables. The static inputs are the network and the link delay functions. The interplay between *demand*, as given by the logit model, and *supply*, as given by the link cost functions, defines the stochastic user equilibrium. The most significant paths in the network are generated by an approximate steepest descent column generation method. To better represent the transitory overloads which characterise congested conditions, a time-dependent PFE has been formulated. In this paper, results from the time-dependent PFE applied to the networks of Lyon, Toulouse and Turin will be presented, and the application of the PFE within the COSMOS project will be outlined. A full description of the PFE plus up-to-date references may be found on **http://www.ncl.ac.uk/~nsg5/PFE/pfe.html**.

2 PFE Theory

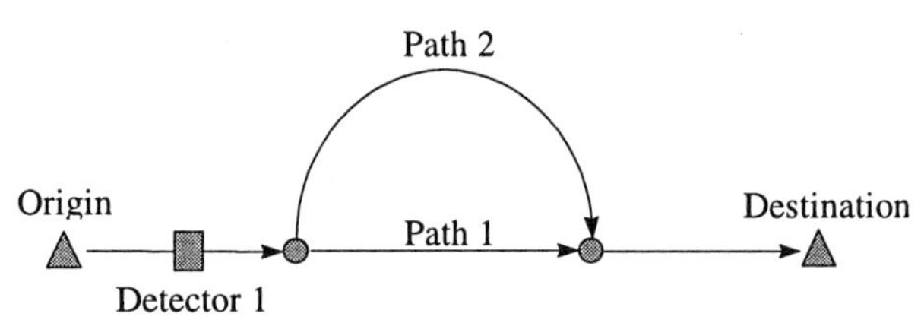

Figure 1: A simple network.

Let us consider the two-path network in Fig. 1. From a measurement of flow made at the detector indicated on the approach to the bifurcation, the problem is to estimate the flows on paths 1 and 2. Path 1 appears to be shorter, and might therefore be expected to attract more traffic.

The relationship between the cost of the two paths and the share of the traffic attracted is determined by the logit response surface (see Fig. 2). The least cost path attracts the majority of the traffic. Link costs are decomposed into two components, a fixed component corresponding to the undelayed travel time and a demand-determined variable component corresponding to the delay. Generally, it is reasonable to suppose that link cost does not decrease with increasing flow.

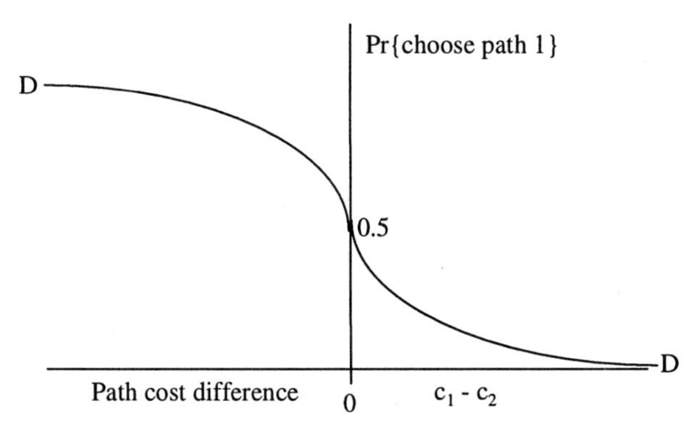

Figure 2: The logit response surface.

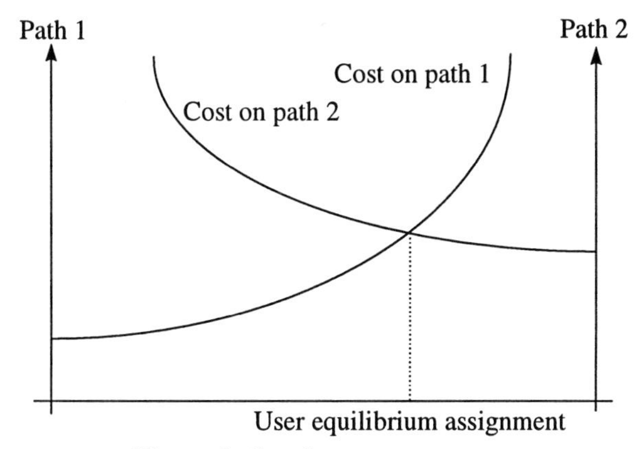

Figure 3: Steady state cost curves.

Fig. 3 shows how the two path cost functions might appear when plotted back top back. To the right of the intersection of the two curves, path 1 is costlier than path 2, while to the left of the intersection, path 2 is costlier than path 1. Under a deterministic user equilibrium assignment, the share of traffic assigned to the two paths would be given by the point of intersection of the two curves.

When the difference in path cost between the two paths is plotted against the share of traffic assigned to path 1, the solution yielded by the PFE is that suggested by the intersection of the two curves in Fig. 4. The curve SS represents the supply-side relationship between the share of the traffic on the two paths and the resulting difference in path cost. This relationship emanates directly from the path cost functions portrayed in Fig. 3. The curve DD represents the demand-side relationship between the difference in path costs and the share of traffic choosing to use the two paths. This emanates from the logit response surface as shown in Fig. 2.

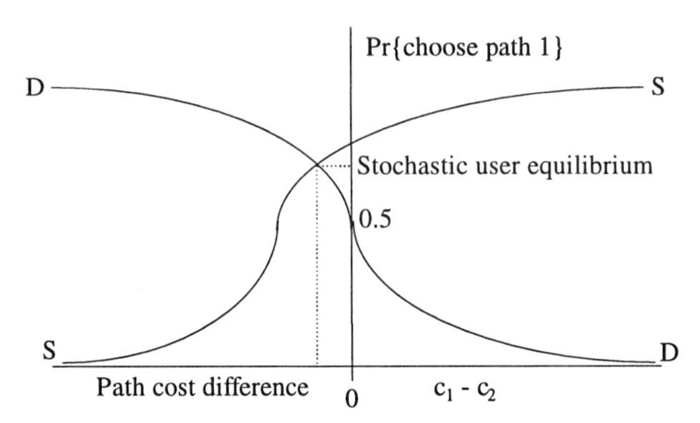

Figure 4: Equilibrium between demand and supply.

The PFE was originally designed to be applied in a steady state environment that presupposes that all trips are completed within the time slice. However, this does not describe congested networks well, since transitory overloads exist and are accommodated by the growth of queues which later discharge when the transitory overload has subsided. A time-dependent PFE has been formulated [2] that assumes steady-state conditions within each time slice but allows queues accumulated in one time slice to be carried across to the next. This useful compromise, which follows that used in the SATURN simulation model [3], allows convex programming method to be retained.

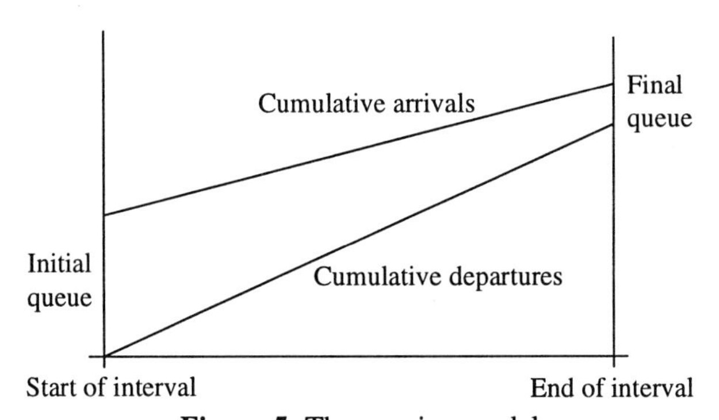

Figure 5: The queuing model

Time slices of between 5 and 15 minutes are considered. Equilibrium delay within each time slice is converted into an equilibrium queue, which is then carried forward to the next time slice. As Fig. 5 suggests, the link queue at the end of one time slice represents a reduction in the available capacity in the next time slice as it will be processed first, thereby conforming to a First In/First Out ordering, at least regarding the processing of the queues.

There is an inconsistency underlying the PFE approach to time-dependency. The assumption of steady-state conditions within the time slice presupposes that all trips are completed, whereas the queues carried across from one time slice to the next consist of incomplete trips. The significance of the inconsistency will depend on many factors, including the size of the time slice used, but is not believed to be great. Thought is now being given to correcting for potential double counting caused by vehicles being simultaneously in a queue and on the downstream links.

3 Path Flow Estimator

3.1 Model

The variables used in presenting the PFE are set out inTable 1. The algorithm builds paths as the iterations progress. The paths are stored in the link-path incidence matrix as columns of elements 1 or 0, indicating the inclusion or exclusion of a link from a path. Let d_{ij} be an element of the link-path incidence matrix corresponding to link i and path j. This will have value 1 if link i is included in path j, and 0 otherwise. Thus the path flows may be summed to yield the fitted link flows as follows:

$$v_i = \sum_{j \in P} \delta_{ij} h_j \quad i \in U \cup C.$$

Confidence intervals defined by the measurements are specified by the user. For the counted links, convergence is reached when:

$$(1-x)\, m_i \le v_i \le (1+x)\, m_i \quad i \in C.$$

where $0 \le x \le 1$ defines the level of confidence in the measurement.

The model essentially consists of the product of an OD-specific weight, a deterrence function and a series of factors:

$$h_j = Q_w \exp(-\alpha t_j) \prod_{i \in C} M_i^{\delta_{ij}} \quad j \in P.$$

where Q_w is the weight specific to OD pair w, $\exp(-\alpha t_j)$ is the deterrence function, and M_i is a factor specific to measured link i. The OD-specific weights are either user defined or calculated so that an input OD matrix is reproduced. In the latter case, the role and therefore the treatment of Q_w is the same as that for M_i.

The dispersion parameter α governs the sensitivity to path cost. When α is large, path choice is sensitive to path cost, whereas when α is small, trips are spread evenly across all the paths with little regard to cost. The M_i and Q_w factors are calculated so that the fitted link and origin-destination flows lie within the confidence intervals. These factors reflect the upward or downward pressure exerted by the constraints on the path flows. For example, when M_i is larger than unity, constraint i is exerting an upward pressure on the path flows passing through counted link i. Conversely, when M_i is smaller than unity, constraint i is exerting a downward pressure on the path flows passing link i.

Sets	Symbol	Variables	Description
Unmeasured links	U	v_i	Fitted flow for link i
		c_i	Fitted cost for link i
		s_i	Capacity for link i
Measured links	C	v_i	Fitted flow for link i
		c_i	Measured cost for link i
		m_i	Measured flow for link i
		M_i	Factor for measured link i
Paths, paths between OD pair w	P, P(w)	h_j	Fitted flow for path j
		t_j	Fitted cost for path j
Feasible pairs of ODs	W	OD_w	Measured flow for OD pair w
		q_w	Fitted flow for OD pair w
		Q_w	Factor for OD pair w

Table 1: PFE sets and variables.

3.2 Link Cost Functions

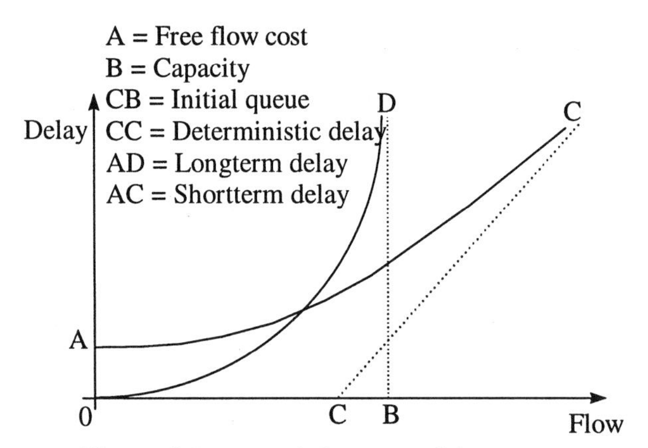

Figure 6: Long- and short-term delay curves.

Fundamental to the PFE are the link cost functions. As mentioned earlier, these are composed of undelayed travel time, which is constant, and delay, which is a function of flow. In a steady state environment, costs would be expected to tend to infinity as the capacity of a link is approached (see Fig. 6). However, this is not what happens in congested networks, because transitory overloads are absorbed as queues which are dissipated later. Hence, the cost would not tend to infinity near capacity.

The relationship between link flow and link cost will also depend on the form of control at the downstream end of the link. The CONTRAM simulator [4] distinguishes between three kinds of link; *priority links* where the capacity depends on geometric factors alone, *non-priority links* where the capacity depends on geometric factors and conflicting priority flows, and *signal controlled links* where the capacity depends on geometric factors and the signal timings. This distinction is also made in the PFE.

Kimber and Hollis [5] present general expressions for time-dependent final (residual) queues and delays. The service process for each link can be regular (as in the case of traffic signal control), random (as in the case of non-priority stream), or somewhere in between (as in the case of the priority link).

3.3 Algorithm

The fitting of the PFE to given static and dynamic data sets is iterative, with an inner and an outer loop. In the inner loop, the path flows are sequentially scaled so that the constraints are fulfilled. The process is iterative, because when the path flows are scaled to conform to one constraint they may no longer conform to the others. This scaling process cycles round all the constraints until convergence is achieved. It can be shown see [2] and [6] that if a feasible solution exists, convergence is assured. In the outer loop, reduced link costs (link costs plus any constraint-induced shadow prices) are calculated and least reduced cost paths are sought. Outer iterations continue until no new paths are generated and link flows cease to change very much. As explained earlier, the OD-specific term Q_w can be either a weight input by the user or a constraint obtained when an accurate estimate (or measure) of the OD matrix is available. The steps of the algorithm are set out in Figure 7.

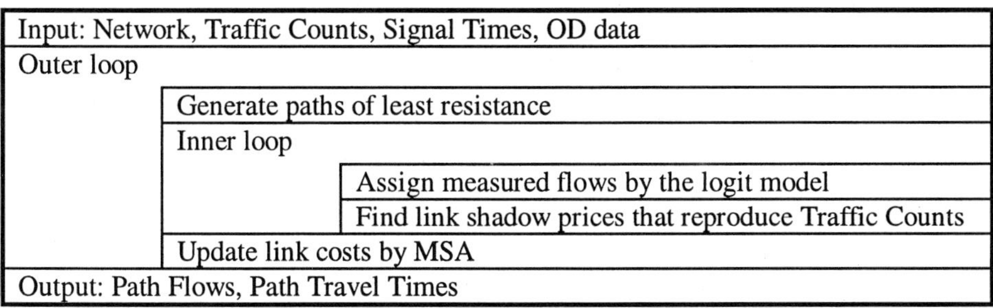

Figure 7: The PFE algorithm.

4 Implementation of the PFE

4.1 Summary of Projects

Project/Sponsor	Objectives
CLEOPATRA	Specification and validation of the PFE for various forms of urban traffic management and control systems in Lyon, Stockholm, Toulouse, and Turin
COSMOS	Specification and validation of the PFE in support of the MOTION traffic signal control system from Siemens and the MAKSIMOS rerouteing optimisation tool from the Technische Universität Hamburg
PTV	Integration of the PFE into the VISION software from PTV GmbH for transportation planning applications
AIUTO	Extension of the PFE to multi-modal networks, with specific reference to the city of York
EPSRC	Use of the PFE to study traffic signal control, with specific reference to the city of York
TASTe	Use of the PFE, as part of a TDM tool box, to study park-and-ride in Barcelona
DETR	Use of the PFE to study traffic diversion due to motorway tolls

Table 2: Implementation of the PFE.

The PFE is scheduled to be tested in a number of European 4th Framework and other research and development projects, as summarised in Table 2. The CLEOPATRA project is concerned primarily with the use of the PFE as a network observer in conjunction with urban traffic management and control. The PFE is being tested in three European cities, where in each case its role is different. In the COSMOS project, the PFE will be supporting the new traffic signal control system MOTION (from Siemens AG) and the rerouteing optimisation tool MAKSIMOS (from the Technische Universität Hamburg-Harburg). The other five projects relate to the use of the PFE for transportation planning, about which subsequent papers will be written.

4.2 PFE Case Study: Turin

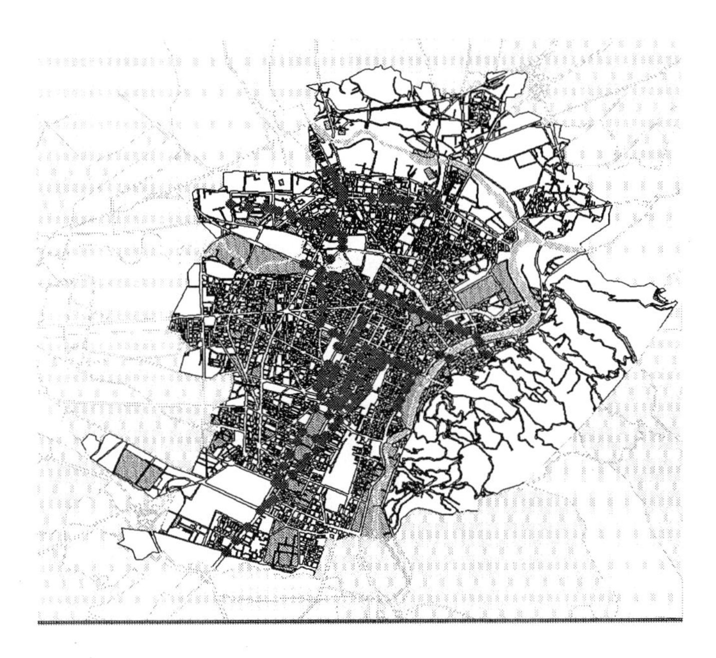

Figure 8: The Turin network.

The Turin network (see Fig. 8) extends over an area of about 1,301,600 m^2, with a total length of 1180 km. The network under the control of the Town Supervisor, with 4530 links of which 2120 are reserved for trams and buses, has been modelled. About 600 links, corresponding to junctions controlled by the UTOPIA traffic control system, are monitored by loop detectors. One week of 5-minute flow data (from 29th September to 5th October 1996) is used to validate the PFE off-line [7].

The validation of the PFE has been carried out by dividing the whole set of traffic flow measurements into two sets, one of them to be used as constraints, the other as test links. 5 test links for each of the two areas under analysis were concentrated on. 7 contiguous days have been chosen, doing the tests for about 144 time slices each day from 7am to 7pm. Tab. 3 in the appendix shows the Mean Absolute Percentage Error (MAPE), and the Mean Absolute Error (MAE) as well as the percentage of estimates to within ±20% of the measured values (%IN) for each day.

Values are also reported for the week as a whole, together with the total number of tests and the number of estimates within the targeted range. Overall, 54.94% of the time the error is within ±20%. The results obtained suggest that in the southern part of the city the measured flows were less accurate than in the northern part, confirmed by the operators of the 5T traffic control centre in Turin. Fig. 9 shows the matching of fitted to measured flows for 6 Corso Unione Sovietica test links chosen at random. The test was carried out using data from 7am to 7pm of 4th October '96. The results could be improved through greater network detail, or by lengthening the duration of the analysis to include 2 or 3 time slices (10 to 15 minutes), while retaining 5-minute intervals for the analysis. To this end, a new network based on the UTOPIA urban traffic control system has been set up to represent the traffic movements at junctions and the PFE has been modified to let the user define the duration of the analysis. Licence plate surveys are planned for one of the two areas to test path flow estimates from the PFE.

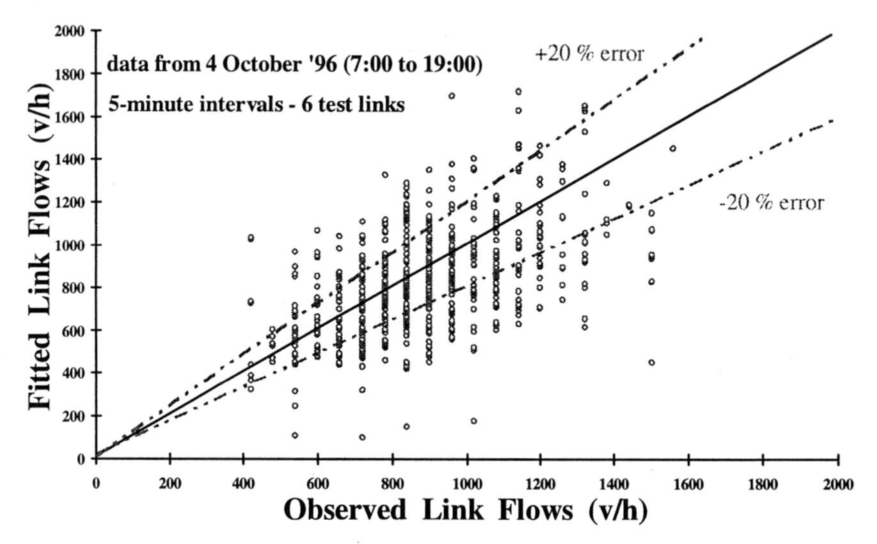

Figure 9: Fitted vs. observed flows for Turin - Corso Unione Sovietica.

4.3 PFE Case Study: Toulouse

The Toulouse network considered in CLEOPATRA is pictured in Fig. 10. The network is composed of 112 links of which about 70 are monitored. As for Turin, the whole set of independent measurements was divided into 2 sub-sets: one are the test links, while the remaining are used as constraints in the PFE. The time slice is 15 minutes. As for Turin, traffic signal data were not available. Tab. 4 in the appendix shows the results of the PFE verification for the period 7th to 13th April 1997, considering 12 hours per day as done for Turin. The overall percentage of estimates to within ±20% of the measured values is 75.10%. Floating car surveys to validate the PFE in Toulouse were planned for April 1998 and the analysis will be carried out by the end of 1998. Fig. 11 shows the result of the estimation for a set of 16 test links chosen at random.

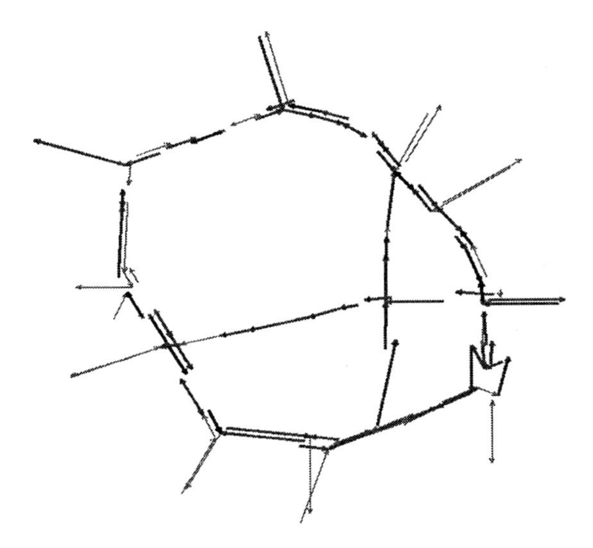

Figure 10: The PFE network for Toulouse.

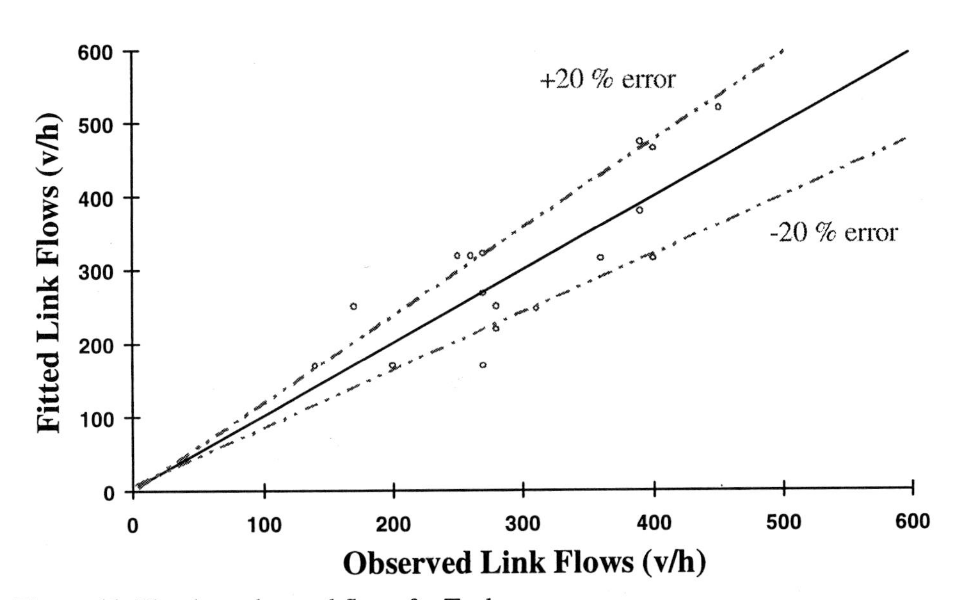

Figure 11: Fitted vs. observed flows for Toulouse.

4.4 PFE Case Study: Lyon

In Lyon, traffic is managed by the „Pascal" traffic control centre. To enlarge the possibilities of control, the Greater Lyon Authorities wish to set up an advanced traffic monitoring system to estimate the status of traffic and predict how it can evolve over a short horizon [8]. The network has 2871 links, 127 of which are monitored. Fig. 12 shows the modelled network.

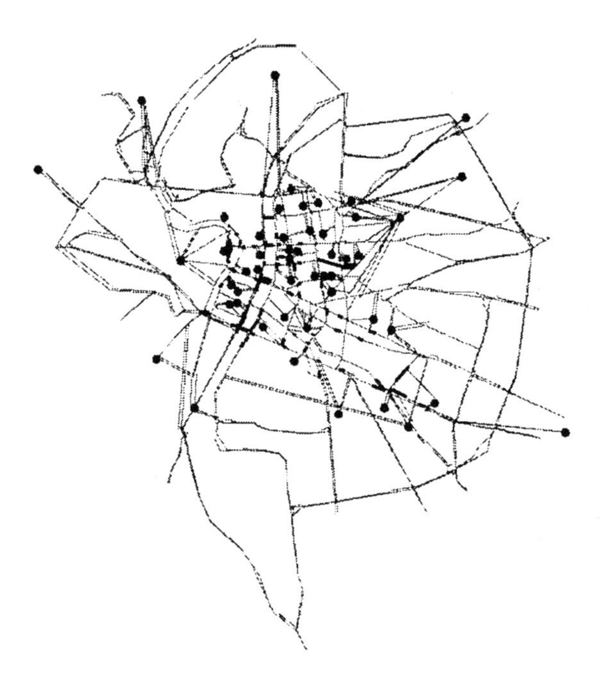

Figure 12: The Lyon network.

Since almost all the links monitored are located in the inner part of the network, it was decided to concentrate on that area. Fig. 13 shows the 699 links (91 of which are monitored) of the reduced network. The time slice is 6 minutes. Since only a small number of links are monitored, we have chosen 6 test links rather than 10, as we did for Turin and Toulouse.

Tab. 5 in the Appendix shows the results of the Verification carried out on the Lyon network for the period from 8th to 14th April 1997. The overall percentage of estimates which are to within ±20% of the measured values is 62.89%. Fig. 14 shows fitted against measured flows for a particular time slice considering 20 test links. Floating car surveys are planned in Lyon to validate the application of the PFE on-line.

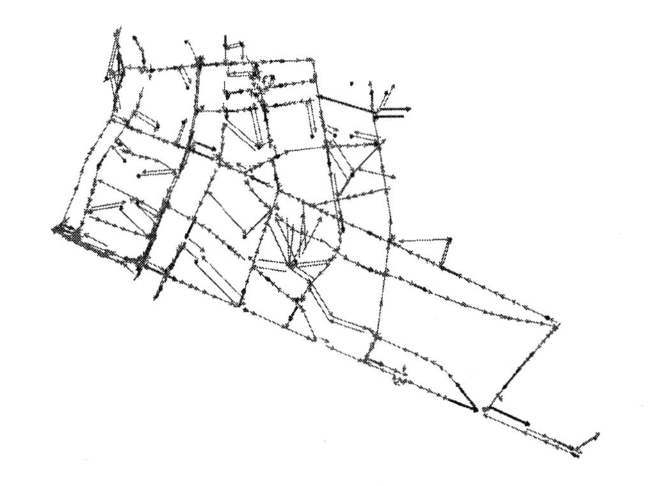

Figure 13: The PFE network for Lyon.

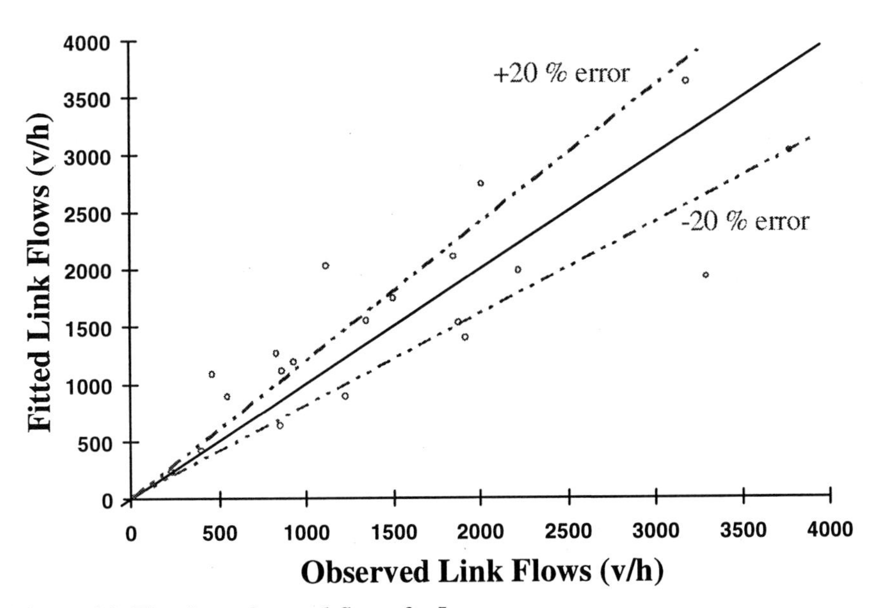

Figure 14: Fitted vs. observed flows for Lyon.

4.5 PFE Case Study: Piraeus

As mentioned earlier, the PFE is being applied in Piraeus within the COSMOS project. COSMOS is a DGXIII European project that will develop, verify and demonstrate new procedures for reducing and preventing congestion in urban areas with high traffic demand [9]. The outputs provided by the PFE are used both by MOTION (from Siemens AG), the signal control system, and MAKSIMOS (from the Technische Universität Hamburg-Harburg), the rerouteing optimisation tool. The main routes for ingoing and outgoing traffic that can be affected by the VMS system are outlined in Fig. 16 and Fig. 17 in the Appendix.

The inner area, where the MOTION traffic control system operates, includes 206 links and 24 junctions. The outer area, which is required by the MAKSIMOS optimisation tool to assess different routeing alternatives for the VMSs, extends the former one by 79 links and 7 junctions. The two networks share the inner part so that the results of the PFE can be exchanged. The actual data come from 68 measurement points (loop detectors) for a total of 35 monitored links. The aggregation interval is 15 minutes. The PFE has been used for two purposes; to assess the general conditions and reliability of the detector outputs, and to predict path travel times. Independent manual counts have been carried out to check the PFE performance on 6 links spread across the network. Fig. 15 shows fitted against measured flows for those surveys performed on September 2nd 1998 from 18:00 to 19:00.

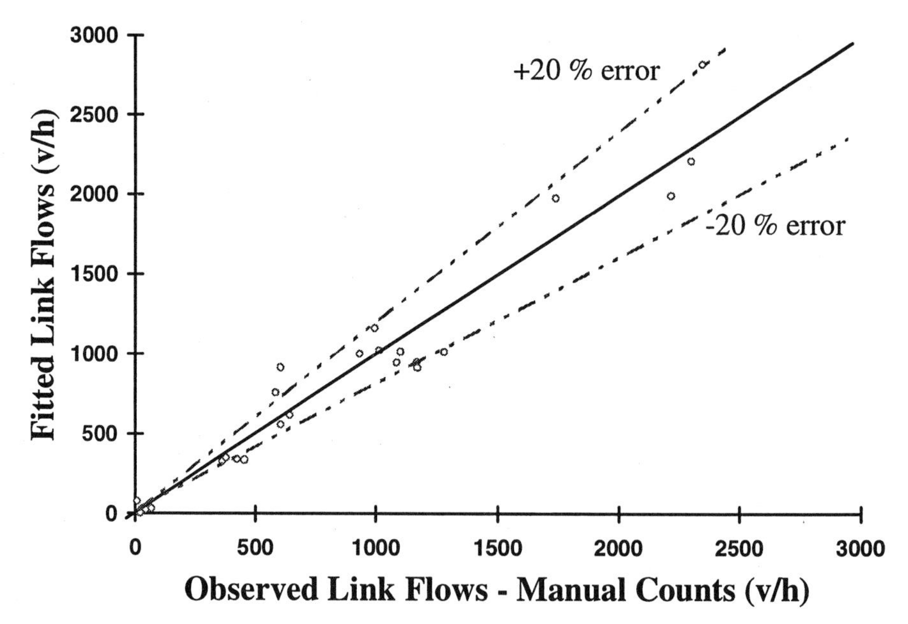

Figure 15: Fitted vs. observed flows for Piraeus.

5 Summary

This paper has presented an one-stage network observer, referred to as the PFE, for use in urban traffic management and control, and has reported initial results obtained in the CLEOPATRA project. The PFE is programmed in C, and can run on a variety of platforms. Efficient use of memory means that large networks can be handled in real time on a PC, although the algorithm stores and processes individual paths. Underlying the PFE is a flexible capacitated stochastic user equilibrium traffic assignment tool which provides unique estimates of path flows as well as travel times. To accommodate transitory overloading, time-dependent delay functions are used and final queues in one time slice are passed forward as initial queues for the next.

Subsequent papers will report on a second version of the PFE which has been developed for use in transportation planning. In contrast to the network observer version, the transportation planning version of the PFE considers multi-modal networks.

Acknowledgments

The authors are grateful to the European Community for supporting the research reported in this paper.

References

1. M.G.H. Bell and C.M. Shield, C.M., *A log-linear model for path flow estimation.* Proceedings of the Fourth International Conference on the Applications of Advanced Technologies in Transportation Engineering, Capri, 695-9, 1995.

2. M.G.H. Bell, W.H.K. Lam, and Y. Iida, *A time-dependent multiclass path flow estimator.* Proceedings of the 13th International Symposium on Transportation and Traffic Theory, Lyon, France, 1996.

3. D. Van Vliet, *SATURN - A modern assignment model.* Traffic Engineering & Control, Vol. 23, 578-581, 1982.

4. D.R. Leonard, J.B. Tough, and P.C. Baguley, *CONTRAM - A traffic assignment model for predicting flows and queues during peak periods.* TRRL Laboratory Report LR841, 1978.

5. R. Kimber and E.M. Hollis, *Traffic queues and delays at road junctions.* TRRL Laboratory Report LR909, 1979.

6. M.G.H. Bell, C.M. Shield, J-J. Henry, and L. Breheret, *A stochastic user equilibrium (SUE) path flow estimator for the DEDALE database in Lyon.* Advanced Methods in Transportation Analysis (L. Bianco and P. Toth, Eds), Springer-Verlag, Berlin, Heidelberg, 75-92, 1996.

7. CLEOPATRA Project TR1012, *Deliverable D4: Final Report on Specification.* DGXIII European Commission, September 1997.

8. L. Breheret, F. Schettini and M.G.H. Bell, *CLEOPATRA Project: Application of traffic monitoring to City of Lyon.* 4th World Congress on Intelligent Transport Systems, Berlin, 1997.

9. COSMOS Project TR1015, *Deliverable D03.3: Integrated UTC Strategies for Congestion and Incident Management.* DGXIII European Commission, October 1997.

Appendix

	Su 29 Sep 96			*Mo 30 Sep 96*			*Tu 1 Oct 96*			*We 2 Oct 96*		
link	MAPE	MAE	% IN	MAPE	MAE	% IN	MAPE	MAE	% IN	MAPE	MAE	% IN
1	20.6	135.4	61.9	15.7	176.1	69	13	145.6	82.5	10.3	128	91.6
2	22.6	103.2	59.7	18.7	153.1	57.1	18	136.1	69.9	16.9	112.8	90.9
3	19.7	117.1	57.6	15.3	151.5	72.4	15.1	145.2	66.1	7.2	49.5	88.8
4	35	102	33.8	26.4	127.6	50.8	26.3	142	36	9.8	48.4	81.9
5	20.2	47.4	58.1	16.6	59.2	67.7	18.1	65.3	59.1	20.4	67.1	20.1
6	47.5	191.3	28.7	38.7	316.6	32.5	47.9	329.9	30.1	0	0	0
7	27.9	125	43.4	32.1	226.8	40.7	28.5	210.5	49.3	20.8	161.4	15.6
8	33.8	166.7	37.5	27.6	274.5	48	32.3	298.2	42.8	10.2	112.6	81.6
9	24.3	108.9	52.7	20.5	179.8	63.3	20.4	174.2	55	34.2	283.6	4.2
10	20.1	71.9	59.7	32.1	222.6	48.8	22.4	182.1	51	43	269	9.7

	Th 3 Oct 96			*Fr 4 Oct 96*			*Sa 5 Oct 96*			*1 week*				
link	MAPE	MAE	% IN	MAPE	MAE	% IN	MAPE	MAE	% IN	MAPE	MAE	in	tot	% IN
1	15.9	146	81.2	15.2	173.9	63	12.6	122.2	81.9	14.7	146.1	743	976	76.1
2	15.6	79.5	85.2	16.3	129.6	73.9	17.2	113.3	65.2	17.9	117.6	706	981	71.9
3	13.3	95.8	82.6	12.2	124.7	79.7	13.3	106.7	81.2	13.7	112.1	736	974	75.5
4	14.4	64.5	78.6	19.8	107.1	49.6	24.7	101.6	44.2	22.3	98.2	506	943	53.6
5	24.6	82.3	25.7	19.8	81.6	52.6	22.1	59.5	50	20.3	66.2	449	952	47.1
6	24.7	163.7	63	29.7	284.1	36.5	28.6	192.6	47.9	36.2	245.6	296	755	39.2
7	85.7	604.3	35.5	19.1	145.6	66.6	17.3	96	69.4	32.9	223.2	446	971	45.9
8	21	181.8	45.5	20.8	184.6	43.4	22.6	170.2	54.8	24	197.3	494	975	50.6
9	32.5	258.7	11.7	25.5	257	27.5	18.6	132.3	54.1	25.1	198.6	374	975	38.3
10	32	209.1	39.4	19.6	158.7	65.4	19.1	116.6	55	26.9	175	452	965	46.8
										Final	result:	5202	9467	54.94 %

MAPE = Mean Absolute Percentage Error
MAE = Mean Absolute Error (v/h)
% IN = Flow estimates (%) to within +- 20% measured values
in = Total flow estimates to within +- 20% measured values
tot = Total flow estimates

Table 3: Results of the PFE Verification in Turin.

link	*Mo 7 Apr 97* MAPE	MAE	%IN	*Tu 8 Apr 97* MAPE	MAE	%IN	*We 9 Apr 97* MAPE	MAE	%IN	*Th 10 Apr 97* MAPE	MAE	%IN
1	26.5	182.3	56.2	15.8	129.4	66.6	13	110.1	81.2	17.6	164.7	64.5
2	19.1	228.4	58.3	14	166.5	81.2	14.9	174.8	77	13.8	166	72.9
3	12.1	193.5	79.1	9.8	157.8	93.7	8.7	146.2	95.8	9.3	152.3	97.9
4	8.8	145.4	93.7	6.1	102	97.9	6.8	115.6	97.9	7.1	117.7	97.9
5	16.5	252.9	66.6	17.5	262.5	58.3	19.3	297.6	52	12.9	186.4	83.3
6	11.6	108.4	81.2	16.9	164.8	66.6	35.7	164.9	68.7	14.1	144.6	72.9
7	15.9	216	72.9	12.3	156.6	81.2	14	181	77	16.9	222.6	75
8	10.3	112.8	89.5	8	89.4	91.6	10.7	104.8	83.3	8.6	88	87.5
9	13	140.4	77	15.6	178.8	66.6	12.5	141.4	81.2	12.9	138.7	75
10	14.8	172.8	66.6	18	216.2	56.2	13.7	170.7	81.2	14.4	171.1	72.9

link	*Fr 11 Apr 97* MAPE	MAE	%IN	*Sa 12 Apr 97* MAPE	MAE	%IN	*Su 13 Apr 97* MAPE	MAE	%IN	*1 week* MAPE	MAE	in	tot	%IN
1	14	113.4	75	13.9	64.7	72.9	43.2	126.4	20.8	20.6	127.3	210	336	62.5
2	12.3	145.4	85.4	12.5	113.5	79.1	16.5	103.6	60.4	14.7	156.9	247	336	73.51
3	11.8	204.3	91.6	9.1	85.5	87.5	12.4	71	85.4	10.5	144.4	303	336	90.18
4	7.9	139.3	91.6	10.3	104.7	89.5	21	148.7	45.8	9.7	124.8	295	336	87.8
5	18.5	267.9	58.3	14.5	145	79.1	19.6	128.8	56.2	17	220.2	218	336	64.88
6	27.4	282.7	52	16.9	129.4	68.7	22.2	85.2	66.6	20.7	154.3	229	336	68.15
7	19	228.1	58.3	13.5	119.3	75	10.7	61.4	85.4	14.6	169.3	252	336	75
8	10.7	113.7	89.5	14	110.7	68.7	9.3	53.2	87.5	10.2	96.1	287	336	85.42
9	13.5	156.4	77	13.3	108.8	75	16.1	101.4	64.5	13.8	138	248	336	73.81
10	16.5	205.2	66.6	16.6	149.3	66.6	12.7	67.9	79.1	15.2	164.7	235	336	69.94
										Final	result:	2524	3360	75.10%

MAPE = Mean Absolute Percentage Error
MAE = Mean Absolute Error (v/h)
%IN = Flow estimates (%) to within +- 20% measured values
in = Total flow estimates to within +- 20% measured values
tot = Total flow estimates

Table 4: Results of the PFE Verification in Toulouse.

link	*Tu 8 Apr 97* MAPE	MAE	% IN	*We 9 Apr 97* MAPE	MAE	% IN	*Th 10 Apr 97* MAPE	MAE	% IN	*Fr 11 Apr 97* MAPE	MAE	% IN
1	15.9	241.6	69.2	19.4	318	60	17	273.9	60.8	17.5	280.1	60.8
2	27.8	399.4	36.7	55.5	490.8	29.2	20.3	288.9	62.5	24.7	365.6	41.7
3	78.7	1232.7	35.8	55.6	917.9	40.8	65.3	1055.2	24.2	32.7	492.5	44.2
4	15.7	280.6	66.1	15.9	285	66.7	17.3	311	55	16.8	303.5	60
5	13.1	415	81.7	16.3	576.7	74.2	11.9	385.3	83.3	9.2	296.3	90.8
6	11	232	85	11	242	83.3	12.6	267.9	75	10.7	229.7	86.7

link	*Sa 12 Apr 97* MAPE	MAE	% IN	*Su 13 Apr 97* MAPE	MAE	% IN	*Mo 14 Apr 97* MAPE	MAE	% IN	*1 week* MAPE	MAE	in	tot	% IN
1	21.8	308.6	49.6	18.6	173	56.7	19.5	264.5	55	18.5	265.6	494	839	58.9
2	27.5	287.3	40.3	20.3	118.2	68.3	11.2	140.1	80	26.8	298.6	430	839	51.3
3	24.7	258.3	58	41.6	159.9	45	27.9	364.6	48.3	46.7	640.6	355	839	42.3
4	13.4	151.8	81.5	15.1	105.5	70	16	258.8	70	15.8	242.3	561	837	67
5	12.1	291.2	82.4	27.8	796.3	42.5	16.3	659.1	66.7	15.2	486.3	617	827	74.6
6	10.3	197.4	89.1	11.9	144.2	82.5	13.9	260.8	85.2	11.6	224.4	693	827	83.8
										Final	result:	3150	5008	62.89%

MAPE = Mean Absolute Percentage Error
MAE = Mean Absolute Error (v/h)
% IN = Flow estimates (%) to within +- 20% measured values
in = Total flow estimates to within +- 20% measured values
tot = Total flow estimates

Table 5: Results of the PFE Verification in Lyon.

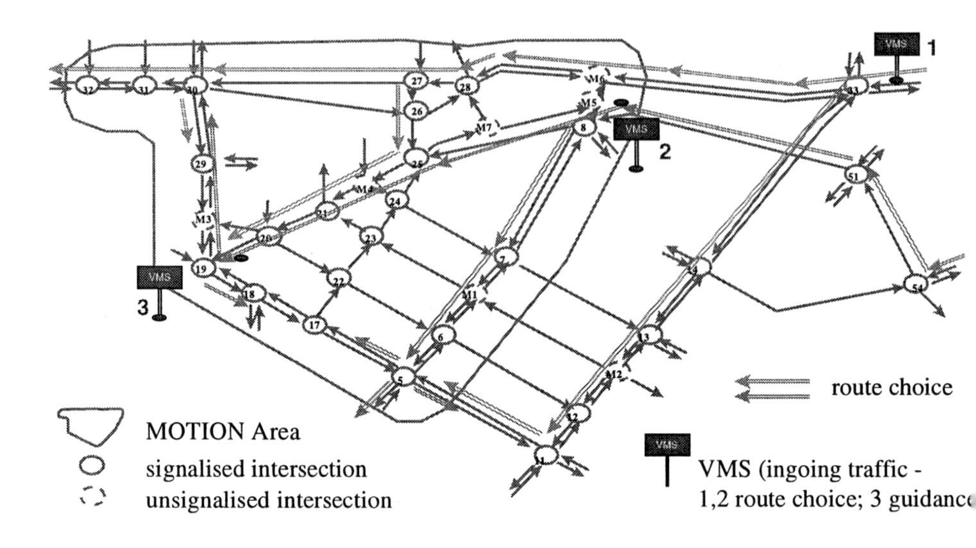

Figure 16: Routeing of ingoing traffic for Piraeus.

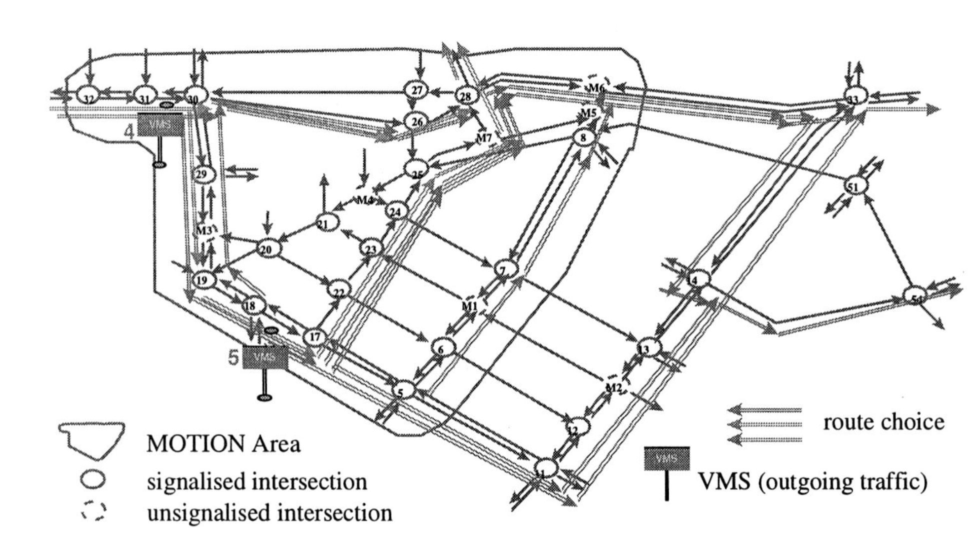

Figure 17: Routeing of outgoing traffic for Piraeus.

III. Traffic Flow Theory, Networks

Remarks on Traffic Flow Modeling and Its Applications

C.F. Daganzo

Department of Civil and Environmental Engineering University of California, Berkeley, CA 94720, USA

This document presents some recent results and ideas from the University of California (Berkeley) traffic operations group, and at the same time discusses the role of traffic flow modeling in traffic management and control. It stresses the steps that can be taken to reduce congestion and improve traffic efficiency, and how traffic models and theories fit within this picture.

1 Introduction

Few drivers would dispute the fact that congestion is caused by bottlenecks, both recurrent and non-recurrent, and that the resulting queues can cause further problems if they become too long. Long queues can entrap cars that do not wish to pass through the bottleneck that generated them, compounding the problem and causing spillovers. These can have widespread effects, such as "gridlock".

It is therefore important to learn more about the behavior of bottlenecks and the spatial extent of queues. The game in congestion management is queue avoidance and containment. This can be achieved through a combination of active control measures such as ramp metering and also by means of passive measures such as route guidance and information delivery.

From a practical point of view, it is most important to have models that can predict reliably the things that matter; i.e. bottleneck behavior and queue dynamics. Models should be tested by verifying how well they can predict these performance measures (the generation of queues and their spatial evolution), more so than other things.

The following text will discuss two important types of bottlenecks (merges and diverges), queue storage issues and the behavior of systems with interconnected bottlenecks. Theoretical and experimental issues will be addressed.

2 Active Bottlenecks

We say that traffic is unqueued ("uncongested" or "free flowing") if small speed disturbances introduced at a location are not detected upstream of that location. Experimental methods have been developed for determining if this is the case [1]. Conversely, if changes in speed introduced at a location are felt later at an upstream location we say that traffic is queued (or congested).

We say that an active bottleneck exists in between two locations if traffic is detected to be queued upstream of the location and unqueued downstream. The identification of bottleneck activity at unexpected locations (non-recurrent congestion) has shown promise for incident detection [2].

3 Merges

A simple theory for merge bottlenecks states that there is a maximum ***sustainable*** flow downstream of a merge when downstream conditions are uncongested, and this quantity is called the "capacity". According to this theory, if the sum of the entering flows over an extended period of time exceeds the merge capacity, then the downstream flow will drop to the capacity level and a queue will form. The queue may grow on either one or on both approaches to the merge, as explained below.

Experiment shows that in some locations the period of time with over-capacity flows lasts for five or ten minutes and at other locations almost nothing at all. Experiments also show that the queue discharge flow (the capacity) is relatively stable. This is most clearly seen from an examination of cumulative count curves [3].

Whether over-capacity flows at merges (if they arise) can be maintained for longer periods of time by doing something to the traffic stream upstream of the merge has not been established. Proper experiments need to control for conditions downstream, which must be uncongested for the duration of the experiment.

Although experiments have not been performed to test the following hypothesis (apparently proposed by K. Moskowitz in the late 1950's) it seems rather plausible. One would expect traffic from the two approaches to a merge to flow in a fixed reproducible ratio if the bottleneck is active and both approaches are queued. Moreover, if one of the approaches is not queued one would expect it to have less flow than its share, and the other approach to take up the slack.

These rules are sufficient to formulate a model that would predict what happens to two traffic streams that compete for space in a merge of insufficient capacity. The model would predict approximately when queues would form and the delay that individual vehicles would experience in passing through the bottleneck. It would not predict the spatial extent of the queues, however [4].

4 Diverges

Diverges, such as the one shown in Fig. 1, are other common locations for active bottlenecks. Here the simplest theory consists of assuming that each off-branch of the diverge has a "capacity" and that flows higher than the capacity cannot be sustained.

If the flow that passes either branch of the diverge exceeds its capacity for a while then a queue should grow in the common approach. If the approach is narrow this queue should operate with a first-in-first-out (FIFO) discipline, so that it will entrap and delay vehicles destined for the other branch of the diverge. Something qualitatively similar may happen even if the approach is wide, as in the case of Fig. 1, but not enough experimentation has been carried out at different locations to determine if this happens in general.

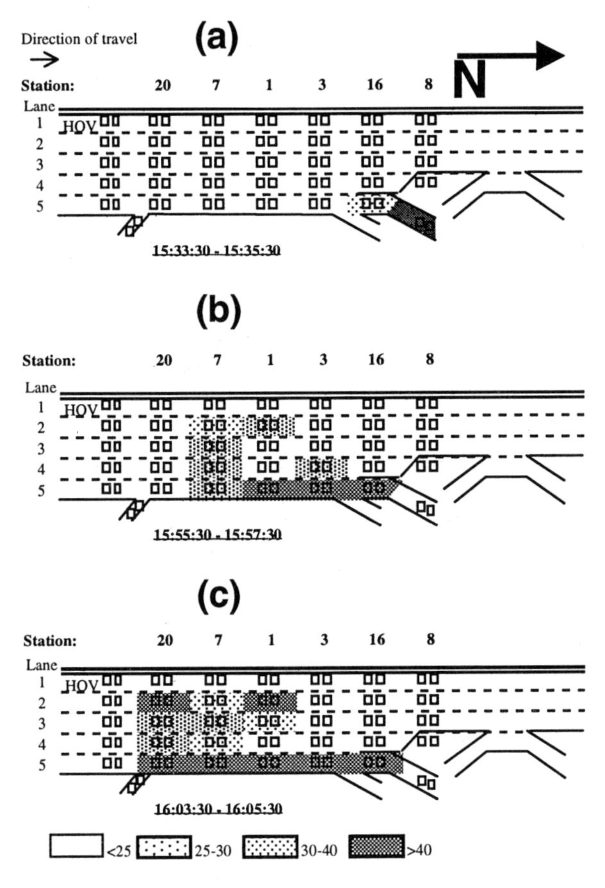

Figure 1: Queue Evolution at a Diverge Bottleneck: Shading intensities indicate occupancy rates [15].

In any case, once a queue has grown upstream of the diverge the combined flow past the bottleneck should depend on the mixture of exiting vehicles present in the queue, which should vary. Thus, ***flows upstream and downstream of queued diverges should not be expected to be steady***, although they could be.

If one knows the desired (virtual) exit flows by destination as a function of time, one can use these bottleneck rules and the FIFO approximation to predict when queues grow upstream of a diverge and the ensuing vehicle delays [5].

A simple geometrical construction with cumulative curves, that can and has been programmed, can be used to illustrate the procedure (see Fig. 2). The example in the figure depicts a curve (labeled "TOTAL") that corresponds to a steady arrival stream.

A queue is created in this stream because the percentage of exiting cars varies. The evolution of delay until dissipation is shown, assuming that each branch of the diverge has a well defined capacity and that the queue discipline is FIFO; i.e., that the horizontal separation between the three pairs of arrival/departure curves is the same for vehicles that arrive at the same time.

Experiments reveal that mainline freeway flows can collapse in this way next to an off-ramp (even if the mainline flow is steady) with no evidence of congestion downstream of the diverge on either branch [6].

5 Other Bottlenecks

Merges and diverges are not the only bottleneck locations. Bottleneck activity can also be detected at weaves, sags, curves, tunnels, lane drops and other locations where the freeway characteristics change. Bottleneck activity can also be generated by temporary exogenous causes such as an incident, some unusual activity next to the road, or even a distracting variable message sign.

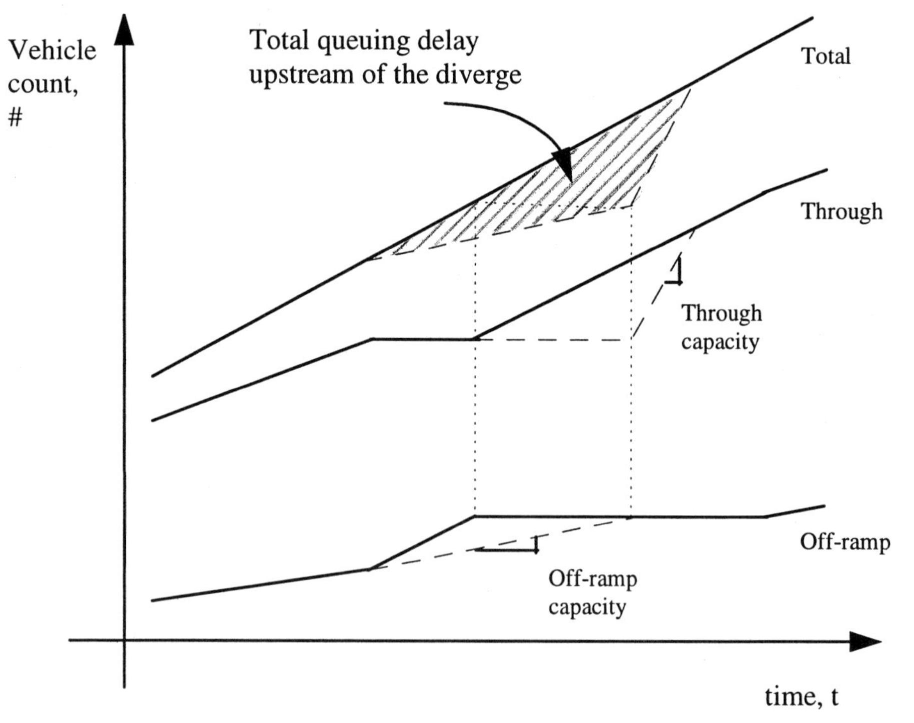

Figure 2: Queue formation upstream of a diverge due to a fluctuation in traffic composition.

In my opinion, it is important to improve our understanding of bottleneck behavior more than almost anything else in traffic theory. Especially since bottlenecks come in many different flavors, and some can be rather peculiar.

As an illustration of this, and perhaps somewhat surprisingly, I should note that there are situations where imposing a speed limit on a freeway bottleneck may actually increase its "capacity". A typical example occurs in a 2-lane uphill section of California State Highway 17 out of Los Gatos, California, where fast cars, which saturate the passing lane at about 110 Km/hr, avoid the shoulder lane because this lane is traveled by just a few but much slower trucks (at 80 Km/hr). The result is a markedly underutilized shoulder lane, which creates a bottleneck at the beginning of the grade. A speed limit could encourage cars to use the right lane, resulting in higher flows through the bottleneck. It could perhaps eliminate the very long waits observed on the approach to the hill.

The above remarks have shown how queues may be triggered (Information of this sort is very valuable because it can be used by traffic management schemes in an attempt to avoid queues). Furthermore, in all the theories reviewed vehicle delays only depend on two things: (1) on people's "schedules", i.e. on the times when individual vehicles would have passed the bottleneck in the absence of queuing, and (2) on the behavior of the bottleneck itself. Nothing else matters. In particular, ***the delay does not depend on the structure of the queues***. This suggests that control actions should be directed at the bottlenecks more than at the queues themselves.

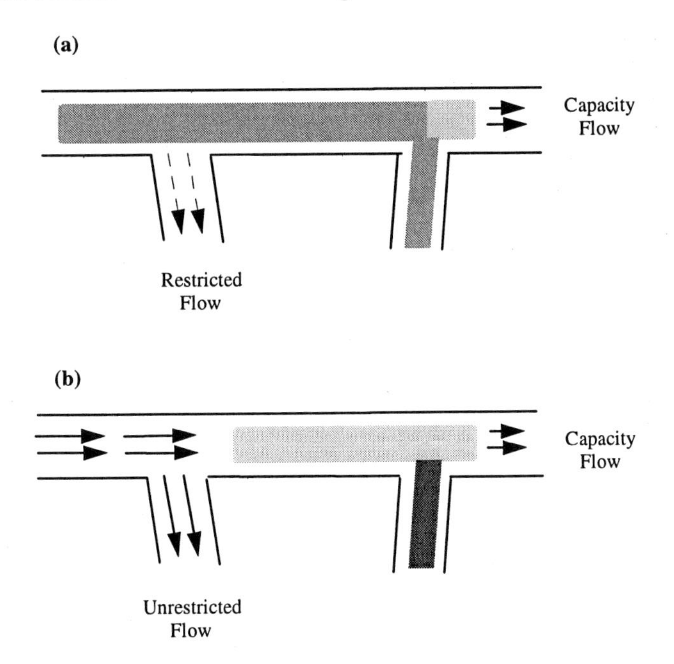

Figure 3: Effect of spillbacks on exit flow and the effect of ramp metering: (a) No Metering; (b) Metering: More System Flow.

This is not to say that the spatial extent of the queues does not matter. If a queue is allowed to spill back past an intersection it may interfere with other ***upstream*** traffic and create another bottleneck. This is usually undesirable.

Gridlock effect: Fig. 3a shows an active merge bottleneck and a freeway queue that spills back past a diverge, starving the exit for flow. According to the foregoing theories if we were to restrict the on-ramp flow, the freeway would take up the slack and the downstream flow would not change. If the ***upstream*** freeway flow could be increased (by metering the on-ramp in this way) beyond the demand level, then the queue would abate. It would no longer block the off-ramp and restrict its flow, as shown in Fig. 3b. As a result the total system flow would have increased. This illustrates the importance of queue containment and the need for theories that will predict the distance spanned by queues [7].

6 Storage Theories

The simplest storage theory would say that the amount of distance per unit vehicle consumed in a long queue (the average spacing) is only a function of the queue discharge rate (Note that this is not very different from saying that there is a "fundamental diagram" between flow and density).

Evidence in this respect is very sparse. Nonetheless, it appears from one experiment [8] (see Fig. 4) that the number of vehicles in between two distinct stationary observers fluctuated within reasonable norms despite the occurrence of stop-and-go episodes, and that this average number increased with declining discharge flows. Thus, the hypothesis made in the first paragraph of this chapter may have some practical value, even if it is not correct at a microscopic level.

This Hypothesis implies the existence of "waves". If carried to its microscopic limit it becomes the kinematic wave (KW) theory. These waves are manifested by linked changes in slope of the curves of vehicle arrival, as suggested in Fig. 5 [9]. In the figure, a downstream change in slope is detected upstream later, which is what one would expect in congested traffic.

Over short distances, waves are noticeable and appear to behave reproducibly [10]. Thus there is hope that this simple model of traffic storage, or a simple modification of it, can predict vehicle accumulations and queue distances reasonably well. In my view, the use of more complicated models for freeway traffic prediction and management seems a bit premature, since it is not clear yet that there is a need.

Macroscopic traffic models that include the above rules for bottlenecks and are consistent with this simple storage theory have been developed. A computer demonstration that illustrates the lasting and widespread effects of an incident downstream of a diverge, and where to some drivers it would appear that the congestion they experienced was caused by nothing, can be seen by visiting the author's home-page:

"www.ce.berkeley.edu/~daganzo/".

A computer program for Windows PC's and its user manual can also be downloaded from this website [11].

7 Spillovers, Information and System Capacity

Spillovers are rather peculiar and they must be treated with caution. Situations exist where provision of information or even the improvement of a bottleneck will actually reduce the flow that can go from point A to point B.

That information can have a negative effect is clear. If we inform drivers on a freeway close to saturation that "there is congestion ahead", some may decide to exit the freeway. If the exit was already close to saturation, the added flow may push the exit "over the edge", converting it into a bottleneck that would spill onto the freeway. In the process we would have created a worse problem than the original one.

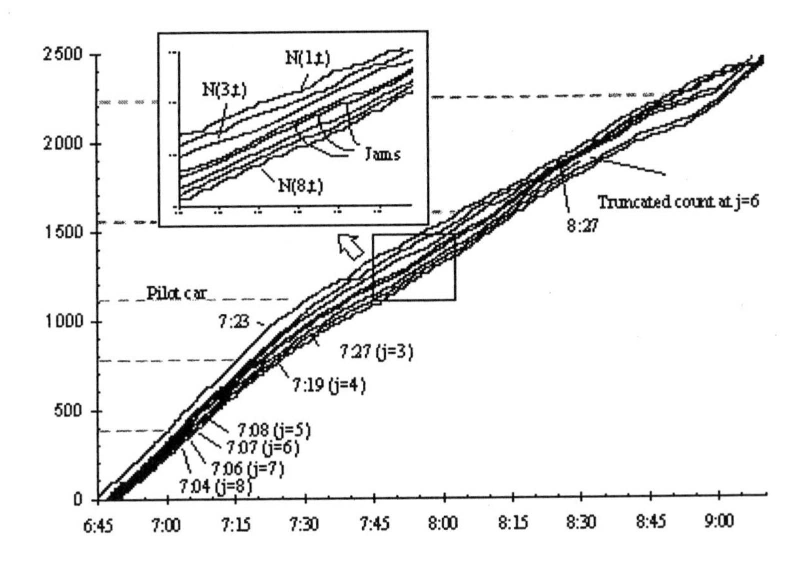

Figure 4: Filtered Curves of N(j,t) for Day Two [16].

Improving an active bottleneck (expanding its capacity) can also be undesirable. This may happen if the bottleneck is located in a preferred road between two points A and B (e.g. a freeway) and the congestion it generates diverts some traffic to less desirable but uncongested alternative routes (e.g. to surface streets). With an improved capacity, the bottleneck should attract enough flow from the alternative routes to equalize approximately the trip times on the preferred and the alternative routes. If the trip time on the uncongested alternative routes is insensitive to decreases in flow, then the delay at the bottleneck should not change either after its capacity is expanded. Thus, its queue will grow... ***longer*** ! The longer queue can spill over upstream intersections. It is not difficult to think of many situations where such a spillover can have dire consequences [12].

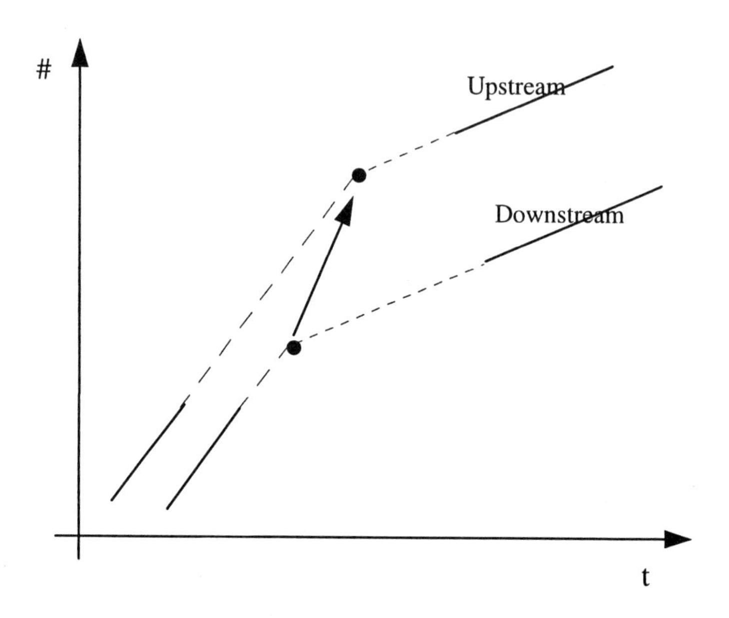

Figure 5: Reproducibility of vehicle accumulations implies a kinematic wave.

8 Multiple Stable States and Chaos

In fact, it is possible to construct very simple networks with steady origin to destination flows, as in Fig. 6, that can be stable in two ways: (a) with an oversaturated configuration where some of the queues have backed up to the origins, blocking their flow, and (b) with an undersaturated configuration of queues that allow all the origins to discharge their flows.

Furthermore, in these types of networks a ***temporary*** disturbance can change ***permanently*** the saturation state of the network. When conditions are changing with time (as occurs in real life), this means that a small localized and temporary variation in conditions can be magnified and spread through the network. The resulting changes, may in turn, trigger others as in a "chain reaction". This suggests that the behavior of traffic in congested networks with interconnected bottlenecks and spillovers is chaotic in nature. This fact, combined with our present inability to predict accurately some basic inputs, such as how people choose routes on a time-dependent network, should force us to rethink the role of predictive models in traffic engineering and management.

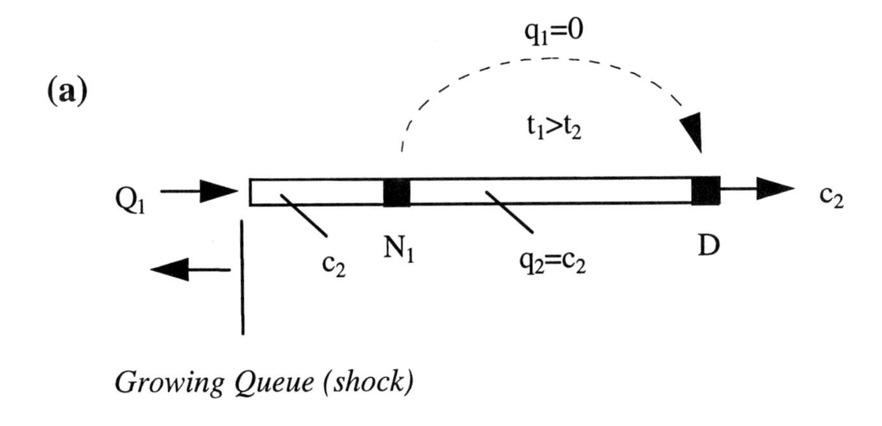

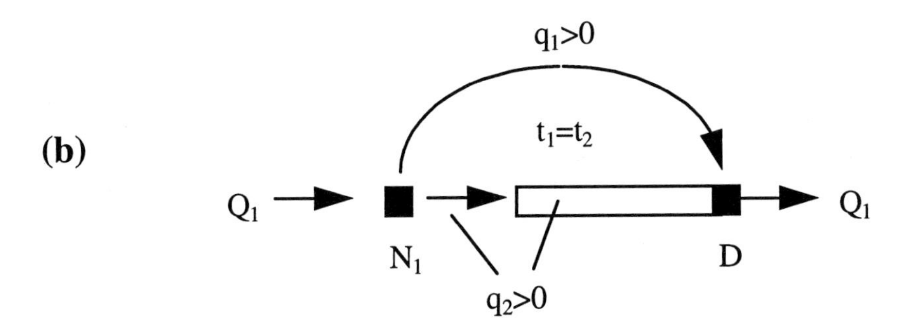

Figure 6: A simple network that can be in two different steady states: (a) Oversaturated; (b) Undersaturated.

9 Role of Modeling

In view of the above, it seems that models cannot be expected to provide precise (or even reasonable) predictions on a link by link basis for large congested networks. Therefore, it seems rather fruitless to base system control procedures (for congested systems) on model predictions.

Nonetheless, models can perhaps play some role in an evaluation process. If one decides for example, that traffic into the central part of a city should be metered (so as to prevent crowding), it may be reasonable to use a ***simple and reasonable*** traffic model to predict the ***overall*** effects of the strategy in the city center. Of course, those predictions would still have to be taken with a "grain of salt".

Simulations are most useful for small networks (or subnetworks such as weaving areas, interchanges and small freeway sections) without route choice [13] where one can hope to obtain the necessary information to make predictions, and where problems of sensitivity of the results to the initial conditions do not arise. In instances like these, methods of dynamic traffic management (e.g. access control and ramp meters) based on models become possible. It is my opinion that in any particular application one should always choose the simplest model that can predict what is at stake [14].

I believe that we stand to gain the most from focusing more attention on two basic questions. Bottleneck behavior is the first one. It is my opinion that too much effort is spent by our research community today developing models of traffic behavior ***for homogeneous roads***, and that not enough experimental work is being done to see how things really are. Homogenous roads are important, but we should really bring the focus of our lenses to the things that matter most. And these are the bottlenecks.

The spatial extent of queues would be the second issue. Experimentalists today measure things such as "occupancy", "speed" and "flow" and use these quantities to assess the goodness of models. I would much prefer to measure the things that the models should really be predicting (e.g. the "queue distances" and "vehicle trip times" between detectors) and then use these to choose among models.

Of course, there are many other issues deserving attention, such as an improved understanding of driver route choice behavior but, in my opinion, the above two research directions are so fundamental that they should be explored first.

Acknowledgments

Much of this work has been supported by PATH MOU-305. The artwork was done by K. Smilowitz.

References

1. One possible approach that seems to be reliable is explained in: "*Methodology for assessing dynamics of freeway traffic flow*, (M. Cassidy and J. Windover), Trans. Res. Rec. **1484**, 73-79, 1995".

2. See "*Incident detection with data from loop surveillance systems: The role of wave analysis*. (W.H. Lin) PhD thesis, Dept of Civil and Environmental Engineering, University of California, Berkeley, 1995." This reference also discusses the state of the art in incident detection.

3. Some interesting figures can be found in "*Some traffic features at freeway bottlenecks*. (M. Cassidy and R. Bertini) Institute of Transportation Studies Research Report ITS-RR-97-07, U. of California, Berkeley, CA, 1997; Trans. Res. A. (in press)," and also in not yet published work by M. Mauch.

4. A detailed explanation of the theory can be found in Sec. 2 of "*The nature of freeway gridlock and how to prevent it*, (C. Daganzo) in *Transportation and Traffic Theory*, pp. 629-646, J.B. Lesort, editor, Pergamon-Elsevier, New York, N.Y., 1996." This reference also describes the implications of the theory for ramp metering.

the above requirements, besides it is perhaps the simplest one. It is capable of reproducing characteristic properties of real traffic, like certain aspects of the flow-density relation and the spatio-temporal evolution of jams (Fig. 1) [7]. Furthermore, *CA* are by design ideal for large-scale computer simulations and can therefore be used in complex practical applications in a very efficient way.

For the sake of completeness, we recall the definition of the Nagel-Schreckenberg *CA* model for single-lane traffic. In the model the street is thought to be subdivided into cells, each *7.5 m* long, which corresponds to the mean frontbumper-frontbumber distance between two consecutive cars captured in a jam (Fig. 2). A cell is either empty or occupied by only one vehicle with a discrete velocity $v_i \in \{0; v_{max}\}$, with v_{max} the maximum velocity. All speeds are measured in *cells per time step*. The motion of the vehicles is determined by the following rules (parallel update):

- Collision-free acceleration: $v_i^{(1)} \leftarrow \min(v_i + 1, v_{max}, g_i)$,
- Randomisation: with a certain probability p do $v_i^{(2)} \leftarrow \max(v_i^{(1)} - 1, 0)$,
- Movement: $x_i \leftarrow x_i + v_i^{(2)}$.

The gap g_i denotes the number of empty cells between two vehicles. A time step corresponds to $\Delta t = 1 sec$, the typical time for a driver to react – an obvious time scale. With a maximum velocity $v_{max} = 5$, for example, the cars can speed up to $135 km/h$. Note that v_{max} does not bound the speed due to technical reasons but is a speed desired by the majority of drivers if they are not hindered by others. In reality, but also in the model, this desired speed is scattered widely. It should also be remarked that a single vehicle in the model might exhibit unrealistic behaviour from a microscopic point of view, e.g. it is possible to slow down from maximum velocity to zero within one time step.

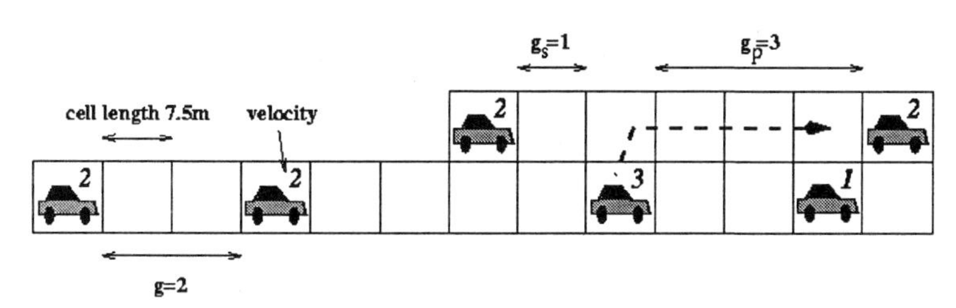

Figure 2: The road in the Nagel-Schreckenberg *CA* model. The road is subdivided in cells. Each car has a discrete speed which is restricted by the headway g to the car ahead and by its own maximum velocity. If a driver wants to change lane he has to take into consideration the gaps g_s and g_p on the alternative lane in order to prevent crashes.

How can these update rules be interpreted? The first rule describes an optimal driving strategy, the driver accelerates if the maximum velocity v_{max} is not reached and brakes to avoid accidents, which are explicitly forbidden. So far, the model is completely deterministic, i.e. the stationary state only depends on the initial conditions. Therefore, the introduction of the noise p is essential for a realistic description of traffic flow. It mimics the complex interactions between the vehicles and is also responsible for spontaneous formation of jams. The parameter p includes over-reactions like heavy braking or delayed accelerations.

2.2 Basic Measurements of Traffic Flow Simulations with *CA´s*

Since *CA´s* describe individual vehicles a lot of information concerning the movement and the environment of every car are accessible. The most common measures are the global quantities *flow J*, *mean speed V* and the *density* ρ. The item *global* is somewhat misleading: on freeways it is only possible to record traffic data on short sections of the road, depending on how long detectors are or how many detectors are installed in a row with a sufficient short distance. Whereas in computer simulations it is easy to determine such measurements. Suppose a simulation is performed on a ring of cells then the density is simply given by the ratio of vehicles and cells *N/L*. If v_i denotes the actual speed of the i-th car, then the main quantities are related via:

$$J = \rho V \qquad with \qquad \rho = \frac{N}{L}, \quad V = \frac{1}{N}\sum_{i=1}^{N} v_i. \tag{1}$$

The dependencies between these quantities are summarised in the fundamental diagram. As an example in Fig. 3 the flow-density relation is depicted obtained from a simulation with the Nagel-Schreckenberg *CA*.

Another interesting entity is the travel time. On the one hand one wants to know how long it takes from the origin to the destination, on the other hand one is interested in the delay caused by jams. In a microscopic model travel times are easy to determine, since every car can be equipped with a stop-watch. Usually, traffic reports are given on the base of estimated jam lengths. But this is not a valuable measure, since jams of the same lengths do not inevitably cause the same delay (e.g. the difference between an obstacle/road blockage and slow-moving platoon of vehicles).

2.3 Extensions of the Model

In fact, more detailed measurements of freeway traffic [8-12] yield that flow is an ambiguous function of density. For some density regimes two branches coexists in the fundamental diagram. The upper branch of higher flow can be characterised by negligible interactions between vehicles and the speeds of the cars is homogeneously distributed, whereas in the lower branch jams emerge. The high-flow states are called *metastable* and have not been observed in simulations with the original Nagel-Schreckenberg *CA* model. But with a velocity-dependent randomisation (*VDR*) Barlovic et al. [13] found metastable states in their simulations. In a *VDR* model the delay probabilities are velocity dependent. With the modification

$$p_0 \equiv p(v=0) >> p(v>0) \equiv p \tag{2}$$

vehicles at the downstream tail of a jam start with reduced probability ('slow-to-start'). The region between two consecutive jams is only rarely filled with vehicles travelling with high speeds and less fluctuations. The outflow from a jam is clearly reduced in comparison to the maximum flow attainable throughout the whole system. This behaviour leads to a modified fundamental diagram (Fig. 3) and a characteristic space-time plot (Fig. 1). After the flow has sharply fallen from the metastable down to the congested branch the traffic state can only return into the free-flow branch by reducing the density clearly.

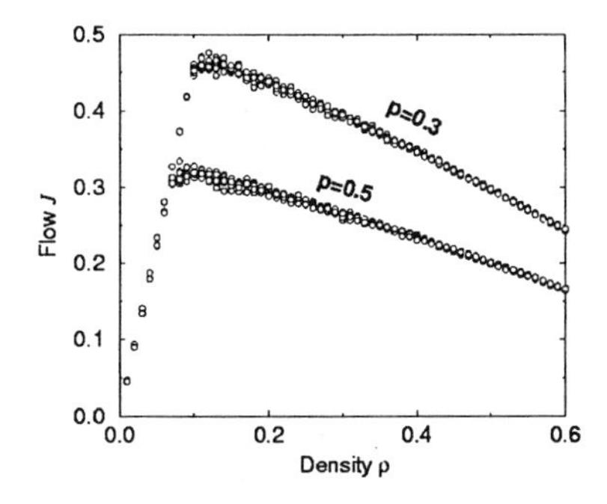

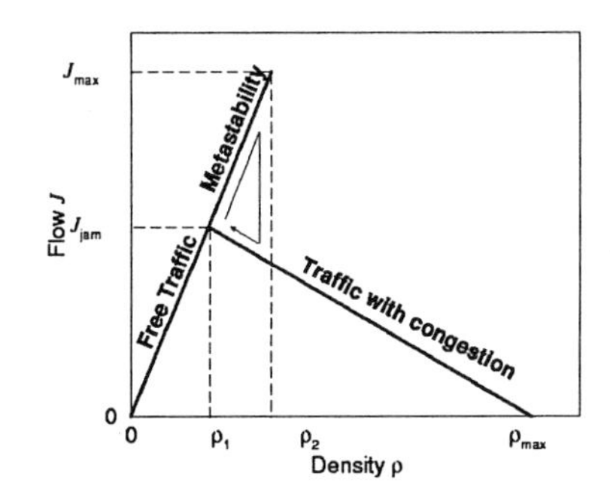

Figure 3: In the fundamental diagram one can see how density controls flow. If less cars are on the road one finds a linear relation until the point of maximum flow is reached. Here the considered system undergoes a cross-over to congested states. Metastability is expressed through the ambiguous flow for a density from ρ_1 until ρ_2. A sudden drop of the flow occurs, after such a sharp fall the density has to be lowered in order to return to the free-flow branch.

A continuous limit of the *CA* which also shows metastable states was earlier proposed by Krauß et al. [14,15]. The effect can be traced back to the finer spatial resolution: In the Nagel-Schreckenberg *CA* model positions and gaps are multiples of the underlying cell length. With a finer grid smaller gaps between two cars are possible, and additionally fluctuations do not have such a great impact since they can be distributed on a wider range of spatial scale and thus be compensated.

Further extensions of the discrete model can help to describe the variety of actual empirical findings, but are still under debate. Promising modifications incorporate aspects of anticipation, realised by taking into account the estimated speeds of predecessors.

Nevertheless, it has been shown that the Nagel-Schreckenberg *CA* model is suitable to model traffic flow in urban networks. In order to describe multi-lane traffic the set of fundamental rules has to be expanded with regard to safety aspects and legal constrains, which vary according to the considered country. A schematic lane change is shown in Fig. 2. Firstly, the vehicle in cell i checks if it is hindered by the predecessor on its actual lane. This is fulfilled if $g_i < v_i$, e.g. Then it has to take into account the gap to the successor g_s and to the predecessor g_p on the alternative lane. If these gaps allow a safe change the vehicle moves to the other lane. A systematic approach for two-lane rules can be found in Nagel et al. [16].

2.4 Analytical Approach

Perhaps the simplest approximation is the mean-field (*MF*) theory [17]. All correlations between adjacent cells are neglected. With $v_{max}=1$ the probabilities c_v to find cars with speed v is expressed through

$$c_0 = (\rho + p(1-\rho))\rho \qquad c_1 = (1-p)\rho(1-\rho). \tag{3}$$

The global density ρ is the sum over all c_i, whereas the current is represented by the number of moving cars: $J(\rho)=c_1$. It is exact in the case of a random sequential update, otherwise the correlations neglected here are of greater relevance (e.g. in a parallel update). Suppose the forbidden states (e.g. two vehicles with $v=1$ after step one of the update rules and a zero gap) are excluded, it yields

$$J(\rho) = \frac{1}{2}\left(1 - \sqrt{1-4(1-p)(1-\rho)\rho}\right). \tag{4}$$

Such "Garden-of-Eden"-states and their effects on the analytical description are described in detail in [18]. In order to obtain reliable results for $v_{max}>1$ the so-called *Cluster Approximation* is appropriated. For this, n sites are combined to a n-cluster, two consecutive cluster overlap each other, e.g. for the case of $v_{max}=2$ it is necessary to set $n=5$ to find a good agreement between analytical and simulation results. The effort grows rapidly with v_{max}, since a system of $(v_{max}+1)^n$ non-linear equations has to be solved, and only large n of the order of magnitude of the range of the vehicle-vehicle correlation are suitable for approximations of good quality.

For the so-called *Car-Oriented Mean-Field Theory COMF* [19] one has to change from the site-oriented to a car-oriented point of view. In the *COMF* the gap-distribution is investigated, i.e. the temporal evolution of the probability to find vehicles with gap n at time t and its distribution.

As an example, the *COMF* method is applied to the *VDR* model [20] (Section 2.2). In this modification two relevant stochastic delay parameters p and p_0 were introduced (2). Beside the gap distribution $d_n^v(t)$ for cars with speed v and headway n at time t one is interested in the probabilities $\gamma_{v(t)v(t+1)}$ for a transition from $v(t)$ to $v(t+1)$. To simplify matters it is defined:

$$q \equiv 1-p \qquad q_0 \equiv 1-p_0. \tag{5}$$

For the case of $v_{max}=1$ the four transition probabilities read

$$\begin{aligned} \gamma_{00} &= d_0^1 + p_0\sum_{n\ge1} d_n^0 & \gamma_{01} &= q_0\sum_{n\ge1} d_n^0 \\ \gamma_{10} &= d_0^1 + p\sum_{n\ge1} d_n^1 & \gamma_{11} &= q\sum_{n\ge1} d_n^1 \end{aligned} \tag{6}$$

which are constrained by a normalisation $\Sigma\gamma_{nm}=1$. Introducing the abbreviations $\gamma_0=\gamma_{00}+\gamma_{10}$ and $\gamma_1=\gamma_{01}+\gamma_{11}$ a system of five *COMF*-equations is considered:

$$\begin{aligned}
d_0^0(t+1) &= \gamma_0\left(d_0^1(t)+d_0^0(t)\right)\\
d_1^0(t+1) &= \gamma_1\left(d_0^1(t)+d_0^0(t)\right)+\gamma_0\left(pd_1^1(t)+p_0d_1^0(t)\right)\\
d_{n>1}^0(t+1) &= \gamma_1\left(pd_{n-1}^1(t)+p_0d_{n-1}^0(t)\right)+\gamma_0\left(pd_n^1(t)+p_0d_n^0(t)\right)\\
d_0^1(t+1) &= \gamma_0\left(qd_1^1(t)+q_0d_1^0(t)\right)\\
d_{n\geq1}^0(t+1) &= \gamma_{11}\left(qd_n^1(t)+q_0d_n^0(t)\right)+\gamma_0\left(qd_{n+1}^1(t)+q_0d_{n+1}^0(t)\right)
\end{aligned} \tag{7}$$

with the condition

$$\sum_{n\geq0}\left(d_n^1(t)+d_n^0(t)\right)=\sum_{n\geq0}\left(d_n^1(t+1)+d_n^0(t+1)\right)=1. \tag{8}$$

A very comfortable way to treat this problem is to introduce the generating functions

$$f_0(z)=\sum_{n=0}^{\infty}d_n^0z^{n+1} \qquad f_1(z)=\sum_{n=0}^{\infty}d_n^1z^{n+1} \tag{9}$$

for standing and moving vehicles. By means of this functions two relations can be rewritten in a simplified form, namely

$$\begin{aligned}
&\sum_{n=0}^{\infty}\left(d_n^0+d_n^1\right)=f_0(1)+f_1(1)=1\\
&\sum_{n=0}^{\infty}\left(d_n^0+d_n^1\right)(n+1)=\frac{\partial f_0}{\partial z}(1)+\frac{\partial f_1}{\partial z}(1)=\frac{1}{\rho}.
\end{aligned} \tag{10}$$

A further function $g(z)=\gamma_0+\gamma_1 z$ enables to eliminate the sum, a pair of coupled functional equations remains:

$$\begin{aligned}
f_0(z)&=g(z)\frac{pf_1(z)-z\left(qd_0^1+q_0d_0^0\right)}{1-p_0g(z)}\\
f_1(z)&=g(z)\frac{q_0f_0(z)-z\left(qd_0^1+q_0d_0^0\right)}{z-qg(z)}.
\end{aligned} \tag{11}$$

After some tedious manipulations the transition probability γ_0 are expressed through

$$\gamma_0=\frac{p+qd_0^1+q_0d_0^0}{p+q_0}\xrightarrow[\text{state}]{\text{stationary}}\frac{d_0^0}{d_0^1+d_0^0}. \tag{12}$$

The probabilities of the several gaps are not independent of each other, it is

$$d_0^0 = \frac{1}{2q_0}\left(\begin{array}{l} q_0 - d_0^1(q+q_0) \pm \\ \sqrt{\begin{array}{l}(d_0^1)^2(p^2-2p-2q_0+1)- \\ \qquad 2q_0 d_0(1+p+q_0)+(q_0)^2\end{array}} \end{array} \right). \tag{13}$$

Finally, one is interested in the fundamental flow-density relation:

$$J(\rho) = \rho\left(q\sum_{n\geq 1} d_n^1 + q_0 \sum_{n\geq 1} d_n^0 \right) = \rho(1-\gamma_0). \tag{14}$$

The desired density ρ can be derived from (10) at the point z=1. Both equations are quadratic, therefore four solutions exist. Not all solutions are of physical relevance (e.g. negative flows or the flow as a descending function of the density in the free-flow regime).

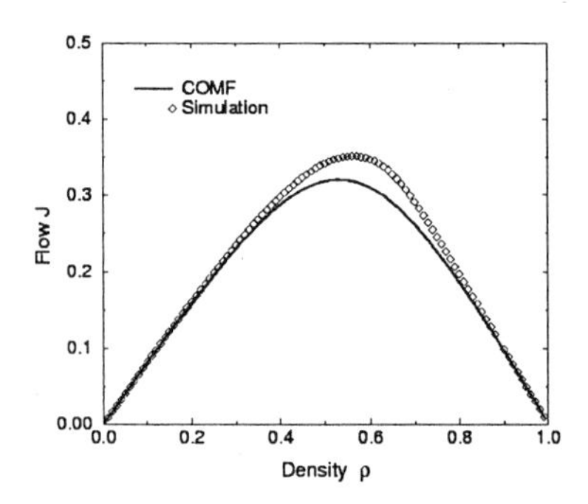

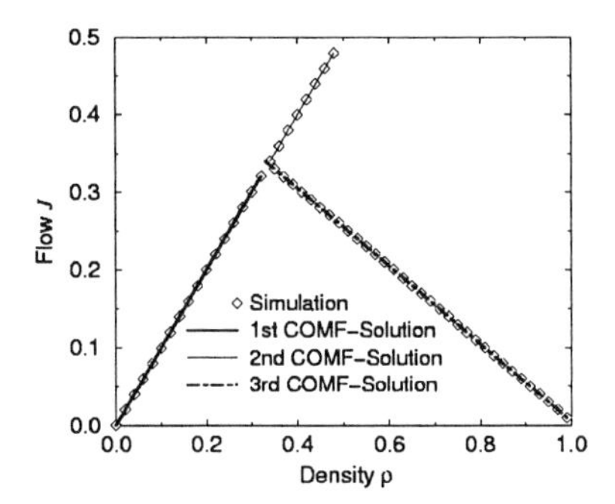

Figure 4: Fundamental diagrams drawn from COMF calculations (v_{max}=1). Left: The point of maximum current is shifted beyond ρ=0.5, which is a hint on downstream moving density waves. The expected particle-hole symmetry known from the original Nagel-Schreckenberg *CA* is broken. Right: In the case of vanishing p the metastability can be reproduced exactly.

In the following paragraph the results of the applied *COMF*-method will be discussed (note that v_{max}=1). For the case p=p_0>0 it yields the expected symmetric fundamental diagram due to the known particle-hole symmetry. As mentioned in previous sections the simulations and *COMF* agree well. In order to make the transition to the *VDR*-model the delay probabilities must differ within in an order of magnitude. Concurrently, the particle-hole symmetry is broken for $p_0 \neq p$, the point of maximum flow (ρ_{max}, J_{max}) is shifted away from ρ_{max}=0.5 (Fig. 4). Surprisingly, for the parameters v_{max}=1, p=0.2 and p_0=0.01 the point of maximum flow is located beyond the density of the half-filled system. This is due to downstream-moving density waves (dense packages of travelling vehicles), any jam dissolves rapidly and therefore they are not stable as long as ρ_{max} has not been reached or exceeded. A similar behaviour can be observed in both the simulations and the *COMF* approach. If the delay probability for driving cars p vanishes and p_0 remains, then two solutions of the *COMF* contribute to the free-flow branch and form the metastable state. The system is completely free of noise until the first jam emerges, then the only stochastic component is the outflow from the jam (Fig. 4). So far it is the only stochastic system without a particle-hole symmetry, for which one is able to derive a *MF* solution.

3 Road Network and Traffic Data

Normally, urban road networks are very complex. But on the other hand Esser and Schreckenberg [21] showed that arbitrary kinds of roads and intersections can be reduced to only a few basic elements. On this basis the complete main road network of downtown area Duisburg, an downtown area of about $30km^2$, was constructed (Fig. 5). In the following an edge corresponds to a driving direction on a road, i.e. each road usually consists of two edges. For each edge the number of lanes, the turning pockets, the traffic signal control and the detailed priority rules are included into the simulation. The network consists of 107 nodes (61 signalised, 22 non-signalised and 24 boundary nodes), 280 edges and 22,059 cells corresponding to about $165km$. The boundary nodes are the sources and sinks of the network.

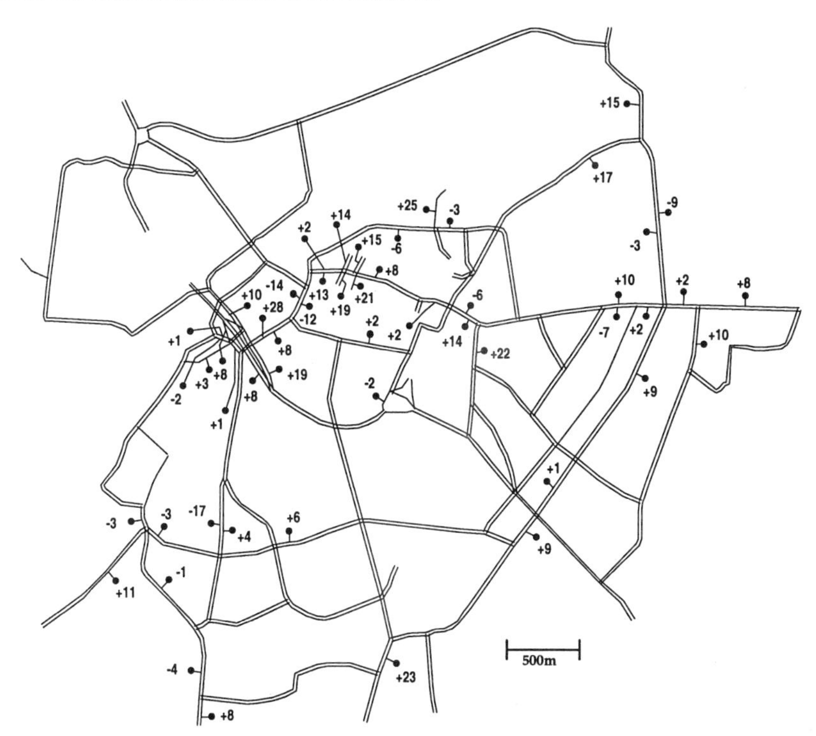

Figure 5: Sketch of the simulated road network with check points (filled circles) and sources and sinks (letters). At the check points the data of all lanes of the road are accessible. Here the local flow can be tuned with respect to the empirical data collection. The digits denote the mean adjustment rates of passing vehicles (cars per minute, averaged over 24 hours).

For an *online* simulation the model has to be supplemented by traffic data gathered from all over the city. Therefore, every minute the measurements at over 730 inductive loops are sent from the traffic computer of the municipality of Duisburg to the online simulation computer (Fig. 6).

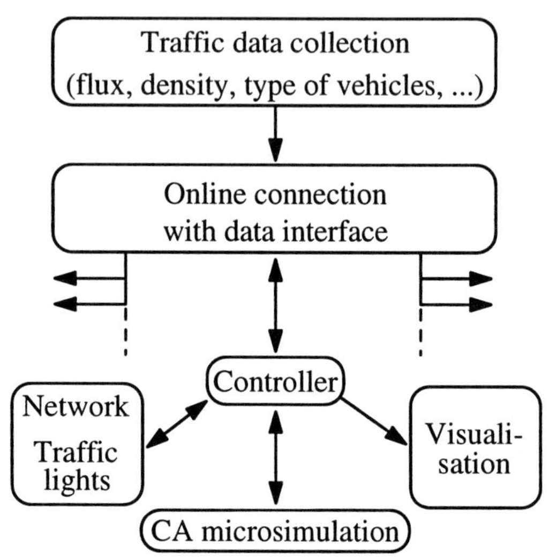

Figure 6: Flowchart of the Online Simulation *OLSIM*. The traffic data are sent via an permanent connection to the Controller, which handles static information like the network structure and performs the *CA* microsimulations. The results can be visualised and processed in further application and numerical experiments.

3.1 Adjustment of the Traffic States

The data collection is used in several ways. Firstly, at the so-called *check points* real traffic data and simulation data are compared. With respect to the differences the number of passing vehicles is changed by creating or annihilating vehicles (internal sources and sinks). It is convenient to perform the adjustments at check points, because at these points (in total there are 51) all lanes of the road belonging to the same direction are equipped with detection devices and therefore cross-section information is accessible (Fig. 5). There are several feasible ways for such recurring adjustments, either by adapting densities resp. mean gaps or flows. To simplify matters the flow-tuning method is used in the described application, but it also turned out, that the quality of the flow-tuning method is sufficient, problems do only occur in high-density states. The following paragraphs are devoted to the explanation of the several methods and a short discussion of their benefits and drawbacks.

Flow-tuning method: The same flow is usually connected with two different densities (Fig. 3), and therefore it may cause problems. But the flow can be easily expressed through the number of cars N_p passing a check point during a simulation period $\Delta\tau$:

$$\varphi_l = \frac{N_p}{\Delta\tau}. \tag{15}$$

An adjustment is simple, but if the measured flow is too high, it may happen that not so much cars can be created as necessary. In this case jams emerge at the location where cars are filled in.

Density-tuning method: There are several ways to estimate the local density. One way is to sum up the duration when vehicles are screening the area of an inductive loop:

$$\rho_l = \frac{1}{\Delta\tau}\sum_{i=1}^{N_p} t_i^{occu} . \tag{16}$$

With the knowledge of the maximum coverage of the road (approx. 140*cars/km*) this relative local density can be directly converted into an absolute one. The second method is motivated by (1) supposed the velocities v_i are accessible:

$$\rho_l = \frac{N_p^2}{\Delta\tau\sum v_i} . \tag{17}$$

With respect to the number N_s of time steps during which vehicles are stopping directly upon an induction loop (only available in simulations) eq. (17) can be extended to a third more precise method:

$$\rho_l = \frac{N_p^2}{\Delta\tau\sum v_i} + \frac{N_s}{\Delta\tau} . \tag{18}$$

The most convenient method, which is applied in this simulations, is the calculation of the density of a complete edge or at least of a section of it: If n vehicles are occupying a link of length z_e, one defines:

$$\rho_l = \frac{n}{z_e} . \tag{19}$$

Tuning the density means to adapt the local density of the simulation to a local density derived from measurements of the real network, in the actual case only given via (18).

In Fig. 5 the rates of vehicles per time interval (typically one minute) that have to be added or erased are shown. Obviously, at most check points cars have to be added into the network due to "internal sources" like smaller side streets or parking lots, which are not recorded by the detection units. This is one main difference between urban and freeway traffic: On freeways sources and sinks are well defined by on- and off-ramps mostly covered by detection devices. So it is easier to handle them and to collect data. Thus, in the urban network of Duisburg reliable results of the online simulation can only be obtained in the downtown area, where the density of check points is sufficiently high.

Furthermore, the collected data are necessary to compute turning probabilities at the intersections, if possible (currently, this can be done for 56 driving directions). In addition, the necessary but missing turning probabilities were counted manually in order to get at least the average number of turning vehicles at intersections which are not covered by measurements. Since a origin-destination matrix with a sufficient resolution in time and space was and is not available, and therefore all vehicles are guided randomly through the network with respect to the given turning probabilities.

4 Results of the Network Simulation

The online simulation enables to interpolate the traffic state between check points (which are typically nearby intersections) and to extrapolate into areas which are rarely or not equipped with detection units. This can be visualised [22] and also serve as a support for planning a trip (in this case for Duisburg, Fig. 7).

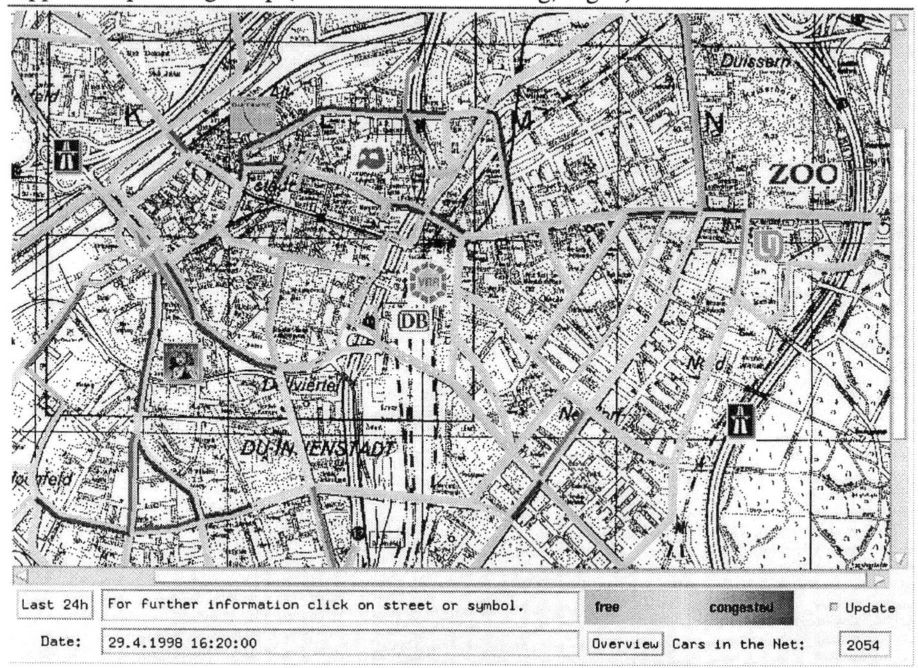

Figure 7: Screen shot of the interactive map [22]. The brightness of the roads is according to the traffic loads computed during the previous minute.

4.1 Typical Days in Network Traffic

The empirical and numerical results allow more detailed examinations of network traffic (Fig. 8). The real traffic data are only collected at selected points of the network, but the simulation extrapolates, therefore one finds more vehicles in the simulation network. On a typical Wednesday (left column in Fig. 8) the commuters cause a pronounced peak in the morning rush-hour. Due to shopping traffic, varying working times and people working overtime the afternoon-peak is broader. Smaller peaks in the early morning or in the late evening are caused by shift-workers. These peaks can also be detected on Sunday (right column in Fig. 8). Additionally, the background traffic is also higher at Saturday night ("Saturday Night Fever").

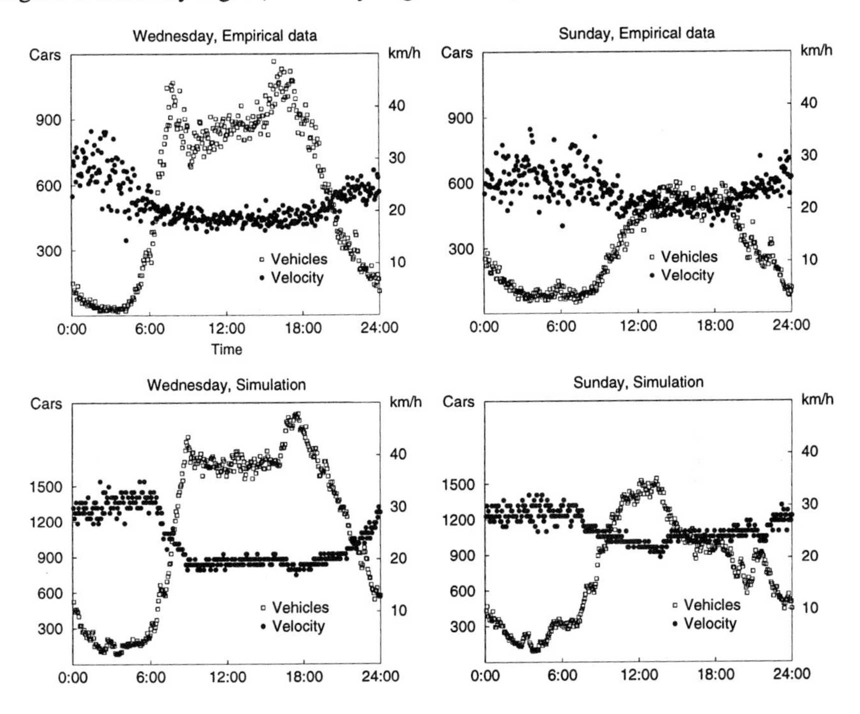

Figure 8: Comparison between the empirical (upper row) and simulation results (lower row) as well as between a typical Wednesday (left column) and a typical Sunday (right column). By night the mean speed is higher than by day, but its variance increases due to the more significant contribution of both cars waiting at intersections and cars driving at high speed on the links.

In general, a smaller number of vehicles driving by night leads to an increased variance of the measured and the simulated speed. Stopping cars at intersections influence the statistics strongly, but it is also highly possible to find drivers who are taking the risk of driving very fast, since the roads are almost empty. By day all these fluctuations are smoothed out. The network is dominated by the intersections and priority rules or their traffic lights, on the roads between intersections the behaviour of drivers is constrained, nearly everyone acts in a similar way irrespective of preferences or engine performance.

The effect of smoothing out can be observed not only between day and night, but also between empirical and simulated results. For the empirical data the drivers' behaviour at the intersections mainly contributes to the statistics, whereas in the simulation the cars travelling between the intersections are of greater relevance. This fact is also responsible for the slightly increased mean velocity.

An important condition for online simulations or, in the next step, generating a *traffic forecast*, is to simulate the traffic in large networks faster than real time. Having in mind the simple set of update rules, *CA* models are by design suitable to meet this requirement. As depicted in Fig. 9, the simulation of a network of the order of magnitude of Duisburg can be easily performed on a common PC 133*MHz*. The simulation includes the data transmission between the diverse modules of the application as well as handling the data of the traffic lights. Within the most interesting density interval (5...20% or 1,000...4,000 vehicles in the network) it only takes a split second to perform an update step of one minute in real time.

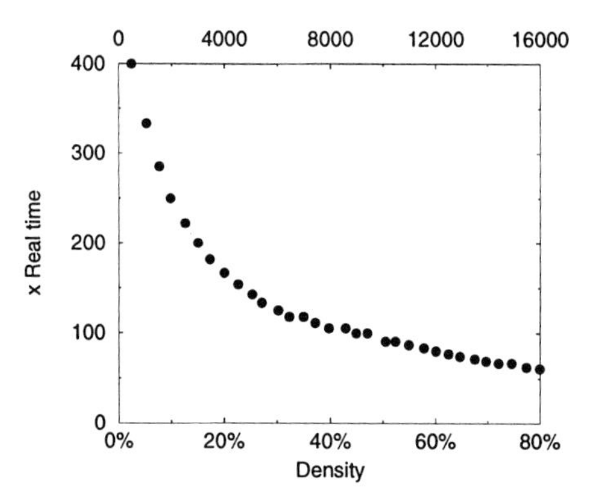

Figure 9: As expected, the simulation speed of a microscopic model like the cellular automaton traffic flow model is inversely proportional to the number of simulated vehicles. For the most frequently occurring densities the road network of Duisburg can be simulated around 100 times faster than real time.

4.2 Reproducing Traffic States

A sensitive test for the quality of the online simulation is the capability to reproduce given traffic states. Because network-wide information cannot be obtained from the measurements, we have to compare the results of the simulations with artificial or *reference* states generated by an independent simulation run. In other words, we perform two simulation runs with two independent sets of random numbers, but the same set of simulation parameters (e.g. source rates, turning probabilities). After reaching the stationary state, we estimated the reproduction rate of the local density via the following quantity:

$$R_\rho = \frac{1}{Z}\sum_{e=1}^{N_e} z_e\,(1-|\rho-\overline{\rho}_e|) \quad with \quad Z=\sum_{e=1}^{N_e} z_e. \tag{20}$$

The edges are weighted by z_e, the number of cells on the edge (hence Z denotes the total number of cells z_e of all N_e edges). The local density of an edge in the second run ρ_e is compared with the local density of the same edge $\overline{\rho}_e$ drawn from the reference state. Each measure point serves as possible check point in the reference run.

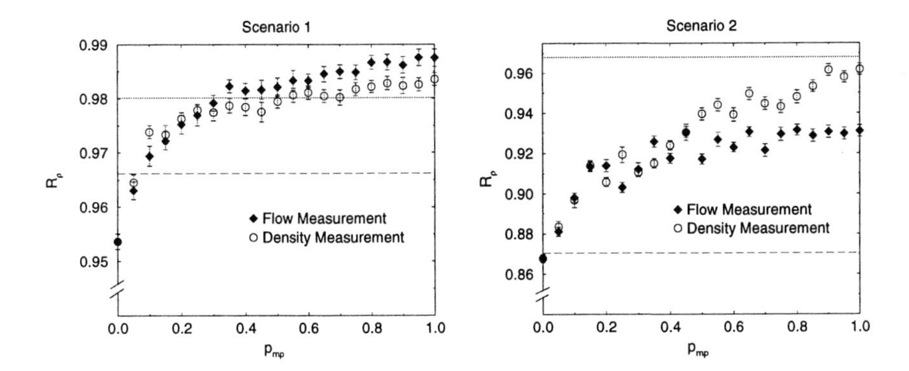

Figure 10: Similarity $R_\rho(p_{mp})$ resulting from reproducing traffic states of two scenarios. Simulations in scenario 1 (or 2) are performed with an input rate r_s=0.1 (or 0.5) at the boundary nodes. Additionally, the special reproduction rates R_ρ^{ran} (dotted line) and R_ρ^{h} (dashed line) are shown.

In Fig. 10 the results of the reproduction rate dependent on the *probability* p_{mp} that a measure point is actually used as check point are depicted. The results are shown for two different values of the input rates at the boundary nodes (scenarios 1 and 2). For obvious reasons the similarity of the states increases for higher p_{mp}. But for small values of p_{mp} the reproduction rate does not increase monotonically. This is due to the fact that the reproduction rate strongly depends on the position of the check points in the network. In order to show this effect we used a completely new check point configuration for each value of p_{mp}. It should be mentioned that even for the empty network (without any check points) high reproduction rates can be obtained if the input rates of the simulation and the reference state agree. Therefore it should be possible to extrapolate future states of the network (*traffic forecast*) with a reasonable accuracy using online simulations.

Comparing both scenarios we can see that the reproduction rate is higher for smaller input rates. This means that it is more difficult to reproduce high density states. Finally, we consider two special cases which are depicted in Fig. 10 as dashed and dotted lines. R_ρ^{ran} denotes the reproduction rate that one obtains if the simulation and the reproduction only differ by the set of the used random numbers, i.e. the measurements are starting with two identical system copies. For large value of p_{mp} a higher reproduction rate than R_ρ^{ran} can be achieved, because the check points also store the fluctuations in the reference run leading to an extremely high value of R_ρ. The reproduction rate R_ρ^{h} denotes the results obtained from an initial homogeneous state in the reference run.

Another criterion for the valuation of the quality of the reproduced traffic state is the time needed for travelling between two nodes. Therefore, we define a parameter D_t that corresponds to the difference of travel times measured in two different simulation runs

$$D_t = \frac{1}{T} \sum_k \left| t_k - \overline{t_k} \right| \quad with \quad T = \sum_k \overline{t_k} \tag{21}$$

as shown in Fig. 11 for two different configurations. The measurements include data collection with *floating cars* (*FC*) which lead to a more sophisticated approach to the reference state.

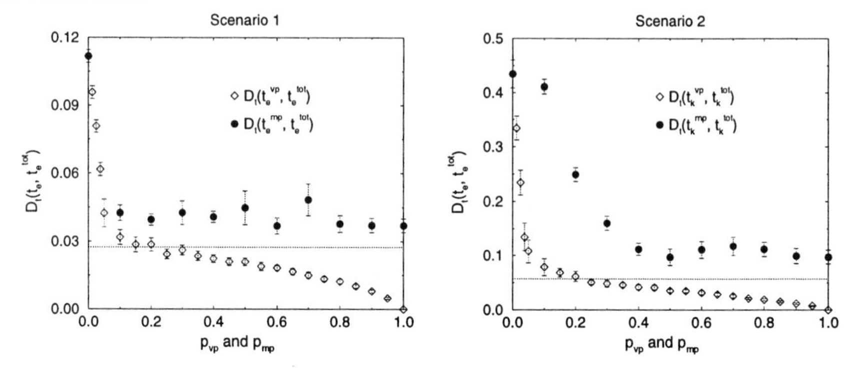

Figure 11: Travel time differences according to (21). The similarity of the compared states increases with the density of floating cars (*FC*) and measurement points (*MP*). The dotted line represents a reference run which only differs in the set of used random numbers, data below of this line characterise a usable reproduction rate.

5 Summary

A simulation tool for urban traffic on the base of cellular automata is presented. The model distinguishes by its simple set of update rules, lane changing or turning at intersections can be implemented in a quite intuitive way and it can be easily implemented on parallel computers. Recent investigations force to modify the original Nagel-Schreckenberg *CA* model in order to recognise the variety of traffic states found in real traffic. Beside the successful simulations analytical approaches have been found and tested. Results of the cluster-*MF* and the car-oriented *MF* theory do agree quite well with the simulation results.

We showed that the Nagel-Schreckenberg *CA* model can be applied to model and simulate traffic flow on freeways and other complex road networks in multiple real time. The combination of high-speed microscopic simulations with real traffic data meets two goals: On the one hand the simulations allow for extrapolating traffic states from a spatial and also a temporal point of view, and on the other hand a useful laboratory environment for designing and evaluating dynamic traffic management systems incorporating different criteria can be implemented.

Concluding, the authors would like to thank J. Lange, acting for the traffic control centre of the municipal authority of Duisburg, for providing essential road network data and especially the online traffic data. We are also thankful to C. Gawron, S. Krauß, K. Nagel, M. Rickert, L. Santen, A. Schadschneider, and P. Wagner for fruitful discussions and sharing insights.

References

1. B. Ran and D. Boyce, *Modeling Dynamic Transportation Networks: An Intelligent Transportation System Oriented Approach*, Springer Berlin Heidelberg (1996).

2. ITS International, *Proceedings of the 5th World Congress on Intelligent Transport Systems*, ITS International Seoul (1998).

3. Avi Ceder (ed.), *Proceedings of the 14th International Symposium on Transportation and Traffic Theory*, Pergamon (1999).

4. D.E. Wolf, M. Schreckenberg, and A. Bachem (eds.), *Traffic and Granular Flow*, World Scientific Singapore (1996).

5. M. Schreckenberg and D.E. Wolf (eds.), *Traffic and Granular Flow '97*, Springer Singapore (1998).

6. K. Nagel and M. Schreckenberg, *A cellular automaton model for freeway traffic*, J. Phys. I, **2**, 2221 (1992).

7. J. Treiterer, *Investigation and measurement of traffic dynamics*, Appx. IX to final Report EES 202-2, Ohio State Univ. Columbus (1965).

8. B. Kerner and H. Rehborn, *Experimental features and characteristics of traffic jams,* Phys. Rev. E, **53**, R1297 (1996).

9. B. Kerner and H. Rehborn, *Experimental properties of phase transitions in traffic flow*, Phys. Rev. Lett., **79**, 4030 (1997).

10. B. Kerner, *Experimental features of self-organization in traffic-flow,* Phys. Rev. E, **81**, 3797 (1998).

11. D. Helbing , *Empirical traffic data and their implications for traffic modelling*, Phys. Rev. E, **55**, R25 (1997).

12. L. Neubert, L. Santen, A. Schadschneider, and M. Schreckenberg, *Single-vehicle data of highway traffic: A Statistical analysis*, cond-mat/9905216 (1999).

13. R. Barlovic, L. Santen, A. Schadschneider, and M. Schreckenberg, *Metastable states in cellular automata for traffic flow*, Eur. Phys. J. B, **5**, 793 (1998).

14. S. Krauß, P. Wagner, and C. Gawron, *The continuous limit of the Nagel-Schreckenberg-Model*, Phys. Rev. E, **54**, 3707 (1996).

15. S. Krauß, P. Wagner, and C. Gawron, *Metastable states in a microscopic model of traffic*, Phys. Rev. E, **55**, 5597 (1997).

16. K. Nagel, D.E. Wolf, P. Wagner, and P. Simon, *Two-lane traffic rules for cellular automata: A systematic approach*, Phys. Rev. E, **58**, 1425 (1998).

17. A. Schadschneider and M. Schreckenberg, *Cellular automaton models and traffic flow,* J. Phys. A, **26**, L679 (1993).

18. A. Schadschneider and M. Schreckenberg, *Garden of Eden states in traffic models,* J. Phys. A, **31**, L225 (1993).

19. A. Schadschneider and M. Schreckenberg, *Car-oriented mean-field theory for traffic flow models,* J. Phys. A, **30**, L69 (1997).

20. R. Barlovic, *Metastabile Zustände im Zellularautomatenmodell für den Straßenverkehr,* Diploma Thesis, Universität Duisburg, Germany (1998).

21. J. Esser and M. Schreckenberg, *Microscopic simulation of urban traffic based on cellular automata,* Int. J. of Mod. Phys. C, **8**, 1025 (1997).

22. OLSIM, Online Simulation Duisburg, Physik von Transport und Verkehr, Universität Duisburg, http://www.traffic.uni-duisburg.de

traffic simulator and the activity generator. Running this on a 250 MHz SUN UltraSparc architecture takes less than one hour computational time for one complete iteration run including activity generation, route planning and running the traffic simulator.

3 Iterative Feedback Experiment

Study Area

Our investigations were are part of ongoing research efforts within the TRANSIMS (TRansportation ANalysis and SIMulation System) project at the Los Alamos National Laboratory. A reduced street network of Portland with 8,564 nodes and 20,024 links, where each link represents one driving direction between two nodes, serves as testing field. For this area a synthetic population of 1,415,900 individuals was generated based on a census from 1990 using an algorithm described in [12]. The resulting household data contains very detailed information (e.g. number of persons, employees, children and cars per household). As mentioned in the introduction, we focus on home to work activities. Thus, only people who work out from home are considered in the following; these account for about 520,000 individuals in the syn-thetic population. Following the purpose of having a minimum model we neglect more detailed individual information (e.g. age, employees' salaries); we do not distinguish different types of employees. In addition to the household data, we use detailed land use data to extract locations and sizes (i.e. number of employees) of companies. For our simulations we make the assumption that all employees work within the study area, since there is no land use and household data available for the surrounding areas. This also means that workplaces can only be found up to a maximum distance. We will refer to this *finite size effect* later when we discuss the workplace distribution.

Simulation Results

Fig. 2 shows how we run the iterative feedback in our experiment: Starting out with a fixed synthetic population, we assign people to workplaces, generate the routes as shortest paths, and run the simulation with these routes from 4am till 12pm. After this, we reroute people four times, while home-work relations are kept constant. Then, workplaces are re-assigned and so on. During the workplace re-assignment, each individual is re-assigned with probability $p_{ra} = 0{,}3$ to avoid over-reactions.

To understand the dynamics of the feedback on a macroscopic scale we first look at the overall travel time (i.e. the sum of all individuals travel times). Fig. 3 shows the total travel time versus the simulation run index (i.e. a sequence of iterative re-assignment according to Fig. 2).

In each workplace assignment step, worker are assigned to workplaces in a way that their expected travel times match the census travel time distribution. Roughly spoken, this means all individuals try do drive an average travel time, which remains constant for all iteration steps, while they face different traffic flow patterns in the network (represented by link travel times). For example, the initial workplace assignment generates too much travel demand, since it is based on the assumption of free speed link travel times. In other words, since initially each driver assumes that the network is empty, the initial workplace assignment generates trips with long geographical distances in order to reproduce the census data. The result of this is a lot of congestion occurring in the first simulation runs, so that in the first workplace re-assignment, individuals are assigned to shorter spatial home-to-work distances to keep the same average travel time. Once individuals are re-assigned, the origin-destination relation for the re-assigned trips and, by this, for the travel demand structure in the network is changed. It takes again some route re-planning steps to adjust routes to the new travel demand structure. In Figs. 4 and 5, snapshots of the simulation are shown for the first and the final simulation run, respectively. The links are colored according to the quotient of average speed and free flow speed; red represents low average speed values. These snapshots show that the overall travel demand is indeed decreased by the re-assignment procedure.

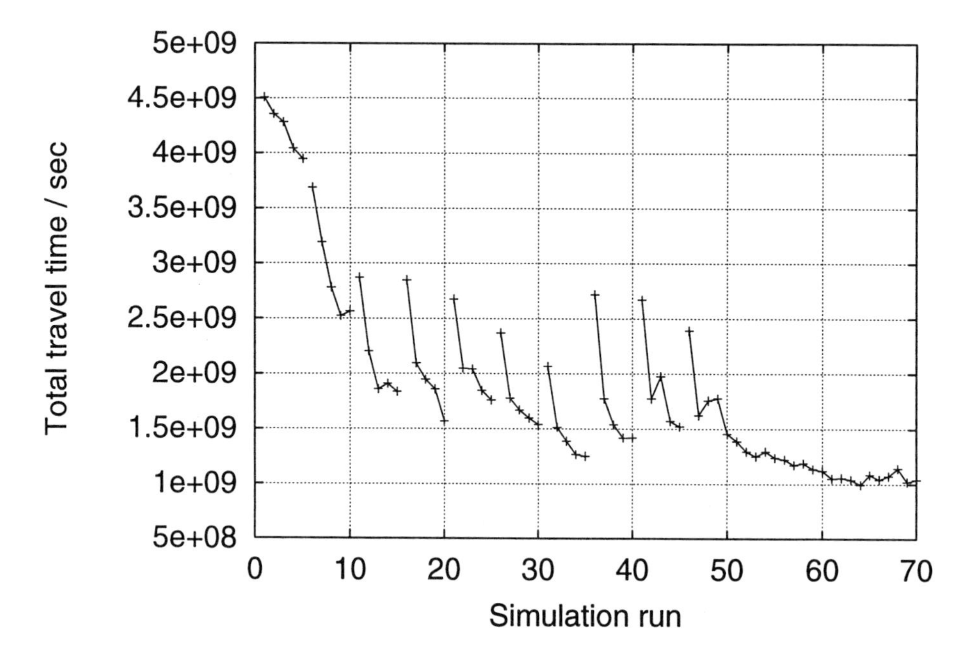

Figure 3: Total travel time in the simulation.

The dynamics of this iterative feedback loop is driven by the probability to find a working opportunity in a given travel time distance. For that reason, it is worth to have a look at the workplace distribution $p_{access}(t)$. Fig. 6 (Top) shows this distribution exemplarily for different activity iteration steps. The initial distribution based on the empty free speed network is linear for travel times up to about 900*sec*. This can be easily understood by considering that the circumference of a circle is proportional to the radius and the workplaces are pretty much homogeneously distributed. For larger travel times, the chance to find workplaces decreases because of the finite size of the study are.

The distributions for later iterations also start out linearly but with a smaller slope, because the average speed on the links is lower than free speed. In addition, the slope fluctuates considerably for those assignments that are based on simulation feedback (i.e. all except the initial assignment). This is due to inhomogeneities with regard to travel times due to capacity constraints in the street network. For example, consider an individual that lives in an area that is connected to the rest of the network only via streets where capacities are exceeded, while within this area no congestion occurs. For this individual, the distribution $p_{access}(t)$ increases steeply for travel time distances within this area. However, once he/she tries to reach workplaces outside this area, it encounters the capacity bottlenecks and has to travel much longer, so there are not many workplaces available at these particular travel times. For this reason, this individual's distribution increases more slowly or even decreases until the first workplace outside the "entrapped area" can be reached. Combinations of configurations like this can lead to plateaus or even local minima in the overall distribution; they reflect typical sizes and distances of isolated (regarding street capacities) regions in the network.

Note that these distributions just reflect the state of the network in the different iterations. Iteration 0 assumes that the network is empty, and thus puts all workplaces within short reach. As a result, iteration 1 is overly congested, and available workplaces are shifted to large times. For all subsequent iterations, traffic gets less and less, which means that the distributions shift back to lower travel times.

Once the working opportunity distribution has been generated based on the last simulation feedback, the travel time acceptance is calculated according to Eq. (4). In Fig. 6 (Middle), the travel time acceptance is shown for different activity iterations.

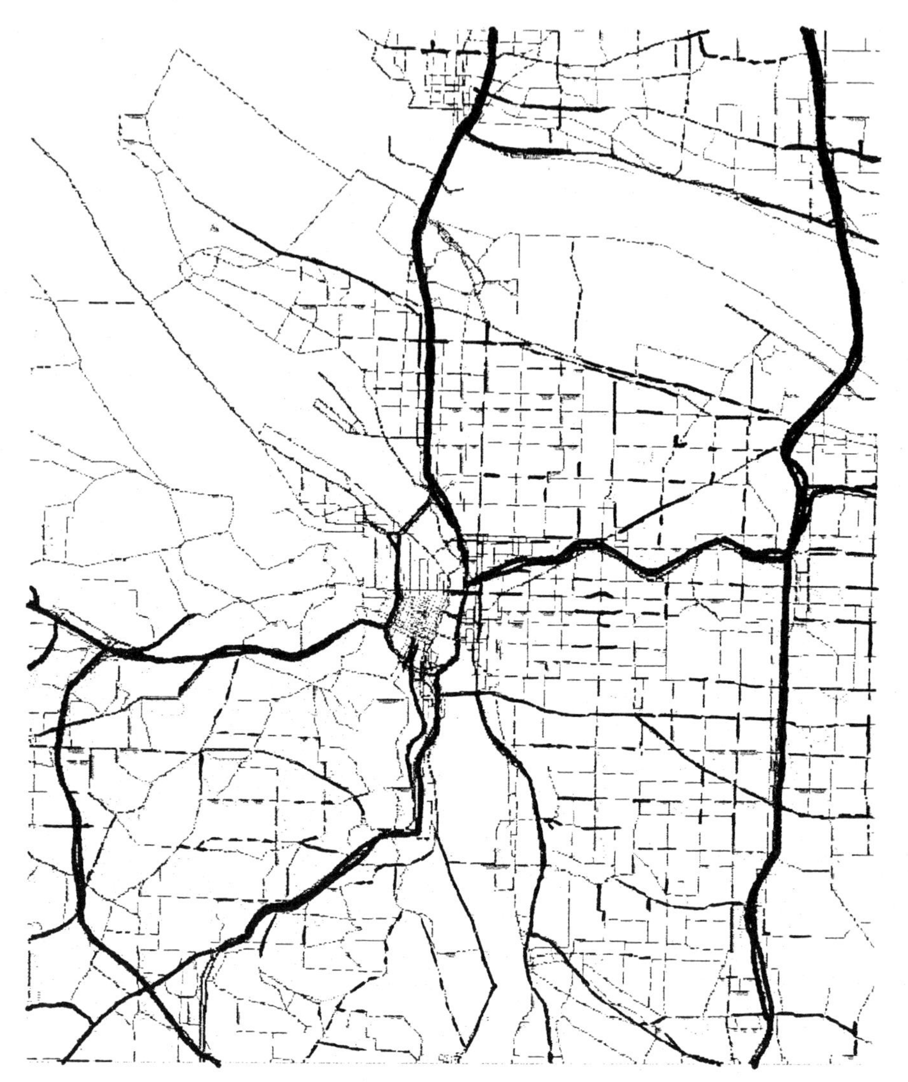

Figure 4: Simulation snapshot at 9am for simulation run 1. Thick lines represent congested roads (speed less than 20% of free speed).

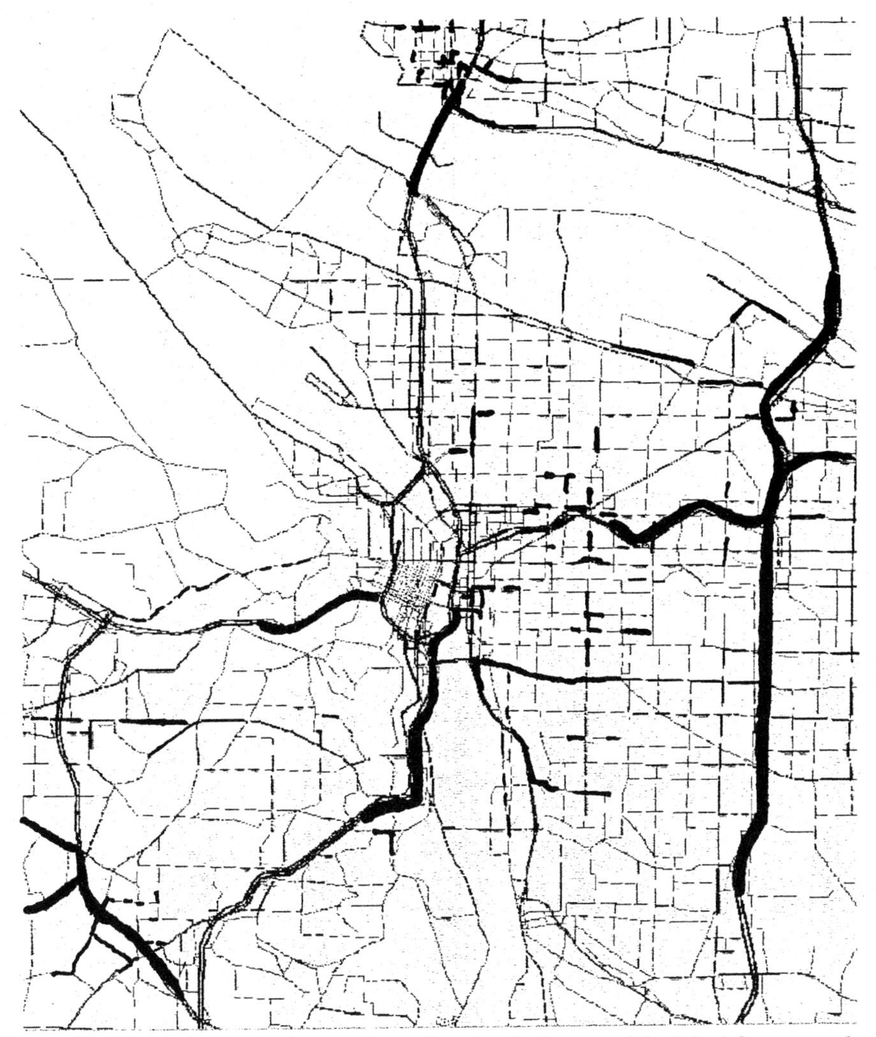

Figure 5: Simulation snapshot at 9am for simulation run 70. Muchfewer roads are congested than in Fig. 4.

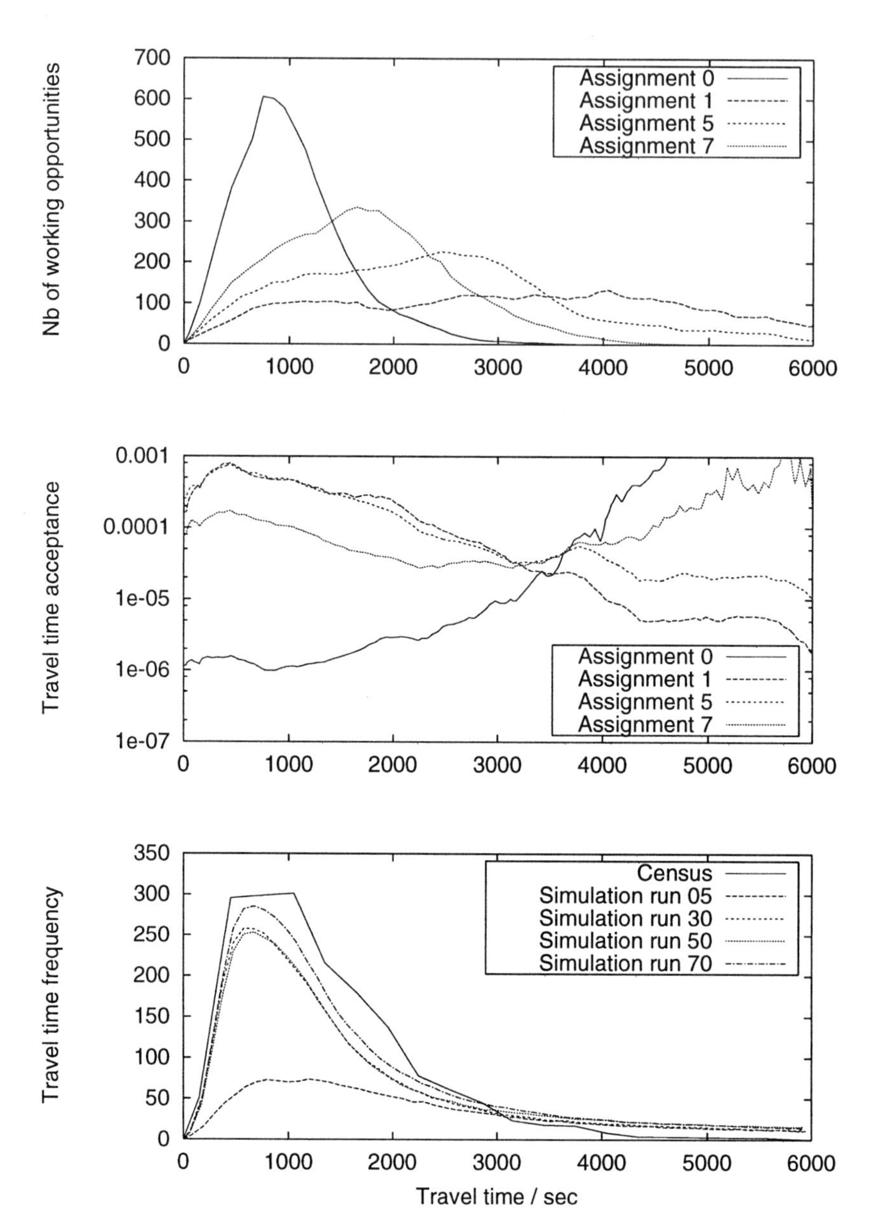

Figure 6: Top: Normalized working opportunity distribution. - *Middle*: Travel time acceptance function $p_{choice}(t)$. - *Bottom*: Travel time distribution resulting from the traffic simulator compared to the census distribution.

First, note that in the initial assignment, one clearly sees the effect of the finite problem size: In our empty and finite network, there are simply not enough opportunities for long trips, and the algorithm compensates by putting a really high probability on these few long trips.

After the first workplace re-assignment ("Assignment 1"), the agents prefer much shorter trips. Since, as we know, assignment 1 is an over-reaction to the initial guess, subsequent assignments allow again for somewhat longer trips. Note that the travel time acceptance for work-place assignment number 7 has again the characteristic that long trips are seemingly much preferred - again, the reason for this is that our finite network does not offer enough opportunities for long trips and the algorithm compensates for that. In reality, working opportunities can be found even for much larger travel times under consideration of workplaces in neighbored cities. Two other arguments should be noted in this context, too: (i) People seem to be more indifferent for long travel times than for medium-length travel times [8]. This certainly would not explain why long trip times should be more preferred than medium-duration trip times, but it would make plausible why there could be a smaller slope in the acceptance function for higher travel times. (ii) It is widely believed that the census over-reports travel times; see below. - All these arguments together mean that there is a variety of possible sources of errors for travel times above an hour (3600 min).

At the other end of the travel times, we also obtain lower preferences for very short travel times (below 5 min = 300 sec). This may indicate that people prefer living in a certain minimum distance to their workplace; maybe simply caused by the fact that people would walk really short distances, thus increasing a driving time of for example one minute to a walking time of five minutes. On the other hand, it may also have something to do with the way we smooth our data for short travel times; see Sec.

Restricting ourselves to the "plausible" range of travel times $400sec < t < 2000sec$, the function $p_{choice}(t)$ can be approximated

$$p_{choice}(t) \sim \exp(\alpha t): \tag{6}$$

Fitting p_{choice} to 6 yields the following values for α:

Re-Assignment	α
1	(-0.0007 ± 0.0003)
5	(-0.0009 ± 0.0003)
7	(-0.0010 ± 0.0003)

Note, and this is really important, that the functional form of the exponential *comes out of the simulations*; it is not invested anywhere in the approach. Thus, what we obtain is another *justification* of ansatz (1), this time not obtained via arguing on the psychological level (as discrete choice theory does), but via making peoples' preferences consistent with their reported travel times in a given transportation network.

Last, we check (Fig. 6 (Bottom)) if the simulated travel time distribution is indeed consistent with the travel time distribution from the census. We see that with higher iteration numbers, the simulated distribution indeed approaches the census distribution, except for too many high travel times. This is somewhat surprising, since in this case one would expect that the acceptance function p_{choice} for "assignment 7" would result in lower weights for long trip times than it actually does. Presumably, the generated travel demands exceed the network capacity in a unrealistic way, and this causes large fluctuations in the travel times of individual travelers [13, 14]. This would mean that always *some* travelers get caught in some heavy traffic, but it is never the same travelers nor the same links, and thus the algorithm in its current form cannot respond to it properly. It is indeed believed that the census overestimates travel times [15]. The explanation for this is that people report the time they *allocate* for the trip, which includes getting ready and walking to the garage, not just the time they are on the road. Also, the simulation model may underestimate network throughput, for example since local streets are missing, and because of the simplifications of the queue model. Resolving this would thus not just mean a precise evaluation (and possible improvement) of our current iteration procedure; it would also involve to calibrate the relation between transportation demand and transportation network throughput. This is beyond the scope of this paper.

4 Discussion

During the re-assignment iterations we keep household locations fixed and changed working locations. This may be realistic for some people, but for other people it might be proper to argue that instead of looking for a workplace in feasible distance, people move closer to workplaces. In our case, the decision was made by the data that was available: Our demographic information is coupled to household locations, and thus the task is to match people and workplaces given the data, *not* to develop a behaviorally entirely plausible model of what people actually do. Future work will hopefully be able to enhance the behavioral aspect of this work.

Looking at the worker/workplace relations resulting from the stochastic assignment, it is necessary to be aware that different microscopic configurations can lead to the same macroscopic travel time distribution. So far, the assignment procedure is entirely driven by overall workplace availability regardless of detailed information As next step to a more detailed approach it is conceivable to impose additional con-straints (e.g. regarding salary) on the assignment procedure. For this, the population as well as the set of available workplaces would be divided into different categories and the stochastic assignment would be applied to each category separately.

Our investigations are confined to home-to-work trips. The underlying dynamics is also applicable for other activity types as far as travel time distributions for those trips are available. Presumably, the acceptance behavior is activity dependent. For example, people accept usually expect longer travel times for their trips to work than for shopping [8]. It would be interesting to see whether the exponential decay in travel time acceptance still holds for other activity types i.e. whether this would only effect the acceptance coefficient α in Eq. 6.

Regarding the iteration sequence, it is possible to combine workplace-assignment and route-planning instead from separating it strictly; i.e. there would be a chance for each individual to change the workplace location in every arbitrary iteration in combination with route-planning. So far, we picked individuals to be re-routed/assigned entirely randomly. It may be possible to speed up the relaxation process by concentrating on "critical" individuals, i.e. individuals that are furthest away from any satisfying choice. For example, one could concentrate on agents where the expected travel time is much different from the travel time experienced in the simulation. In this context, it is also necessary to check in what way the underlying re-routing/assigning algorithm effects the final traffic state after relaxation. Questions related to this are topic of current research.

5 Summary

We presented an iterative activity assignment approach exemplarily for home-to-work trips. The approach allows to systematically generate macroscopic data - for our investigations this was the census travel time distribution - from a transportation simulation. Instead of making any assumptions about the individuals' travel time acceptance behavior, we extract the acceptance function out of the census travel time distribution and the simulation set-up. For this, we decompose the probability of choosing a workplace in a given distance into two terms: The distribution of workplaces in the network, and the macroscopic travel time acceptance function. It turns out that this acceptance behavior can be well described by an exponentially decaying function, which is consistent with other approaches. The fact that the studies were carried out for the real Portland/Oregon road network shows that iterative activity assignment is feasible for realistically sized systems.

6 Acknowledgment

We would like to thank R. Frye, R. Jacob, and P. Simon for a lot of useful hints as well as for developing and maintaining the routing tool and the traffic simulator. We are also thankful to C. Barrett and R. Beckman for many fruitful discussions. Finally, we thank P. Wagner, whose visit at LANL gave an essential incentive to focus on this topic.

References

1. J. Esser and M. Schreckenberg. Microscopic simulation of urban traffic based on cellular-automata. Int. J. of Mod. Phys. C, 8(5):1025-1036, 1997.

2. D. Helbing. Verkehrsdynamik. Springer, Heidelberg, Germany, 1997.

3. B.S. Kerner and P. Konhäuser. Structure and parameters of clusters in traffic flow. Physical Review E, 50(1):54, 1994.

4. R. Barlovic, L. Santen, A. Schadschneider, and M. Schreckenberg. Metastable states in cellular automata. Eur. Phys. J. B, 5(3):793-800, 10 1998.

5. P. Wagner and K. Nagel. Microscopic modeling of travel demand: Approaching the home-to-work problem. Annual Meeting Paper 99 09 19, Transportation Research Board, Washington, D.C., 1998.

6. Y. Sheffi. Urban transportation networks: Equilibrium analysis with mathematical programming methods. Prentice-Hall, Englewood Cliffs, NJ, USA, 1985.

7. M. Ben-Akiva and S.R. Lerman. Discrete choice analysis. The MIT Press, Cambridge, MA, 1985.

8. John L. Bowman. The day activity schedule approach to travel demand analysis. PhD thesis, Massachusetts Institute of Technology, Boston, MA, 1998.

9. P.M. Simon and K. Nagel. Simple queueing model applied to the city of Portland. Annual Meeting Paper 99 12 49, Transportation Research Board, Washington, D.C., 1999.

10. M. Rickert and K. Nagel. Issues of simulation-based route assignment. Los Alamos Unclassified Report (LA- UR) 98-4601, Los Alamos National Laboratory, see www.santafe.edu/˜kai/papers/, 1998.

11. K. Nagel and C.L. Barrett. Using microsimulation feedback for trip adaptation for realistic traffic in Dallas. International Journal of Modern Physics C, 8(3):505-526, 1997.

12. R.J. Beckman, K.A. Baggerly, and M.D. McKay. Creating synthetic base-line populations. Transportation Research Part A - Policy and Practice, 30(6):415-429, 1996.

13. T. Kelly. Driver strategy and traffic system performance. Physica A, 235:407, 1997.

14. B. Raney, J. Esser, M. Hamada, and K. Nagel. In preparation.

15. K. Lawton. Personal communication.

Some New Approaches to the Microscopic Modelling of Traffic Flow and the Dynamic Route Assignment Problem

R. Böning[1], G. Eisenbeiß[2], C. Gawron[1], S. Krauß[2], R. Schrader[1], and P. Wagner[2]

[1]Zentrum für Paralleles Rechnen (ZPR), Weyertal 80, 50931 Köln, Germany

[2]Deutsches Zentrum für Luft- und Raumfahrt (DLR), Linder Höhe, 51147 Köln, Germany

In order to simulate the transportation system of a large region dynamically, three things have to be known and modelled: who wants to go where at what departure time (destination choice), which route to the destination is selected (route choice), and finally how the locomotion along this route is performed in time (travelling). This article deals with the second and third question. Firstly, it is shown how the dynamic route choice problem could be solved by a simulation–based approach. Surprisingly, when comparing this approach to the classic static one, it turns out to be simpler and applicable to larger networks than its classical counterpart. Secondly, by using a simplified model of car–following, most of the phenomena observed in real traffic can be reproduced by the simulation. The realism achieved with this model allows a reliable estimation of the emissions caused by traffic (car traffic amounts to roughly 80 % of the total travel volume). The model is numerically very efficient, enabling the simulation of large road networks with more than 1 million individual cars in real time. Finally, preliminary results are mentioned concerning the possibility of expanding this simplified microscopic approach to the destination choice modelling. Following this route of thinking further finally leads to an integrated, microscopic, and dynamic approach to transportation planning and control. The first steps toward this aim has been done within the FVU.

1 Introduction

Understanding the complexities of a transportation system requires the help of sophisticated computer algorithms. Unfortunately, there is no fundamental equation modelling the dynamic phenomena of traffic observed during a single day which could be solved numerically. Instead, a simulation has to be used in order to describe what is going on in a transportation network. Basically, the research program pursued by the FVU [1] and by similar projects world–wide [2, 3] uses a three-stage approach which resembles the traditional four-stage-algorithm. The three stages (destination choice, route choice and travelling) are coupled through a double iteration loop that computes the state of a transportation system for a characteristic time-period, e.g. for a day. It differs from the four-step-algorithm in using a microscopic modelling wherever possible. At first glance, the microscopic modelling seems to increase complexity. But by taking a closer look it can be seen that it is much simpler to understand a transportation system when thinking microscopically, i.e. in terms of single customers, travellers, cars.

Complication comes in by the fact that the three stages are mutually dependent on each other, e.g. the route choice depends on the traffic conditions which itself depends on the routes chosen. Therefore, the double iteration loop mentioned above is needed, whose principal form is shown in Fig. 1.

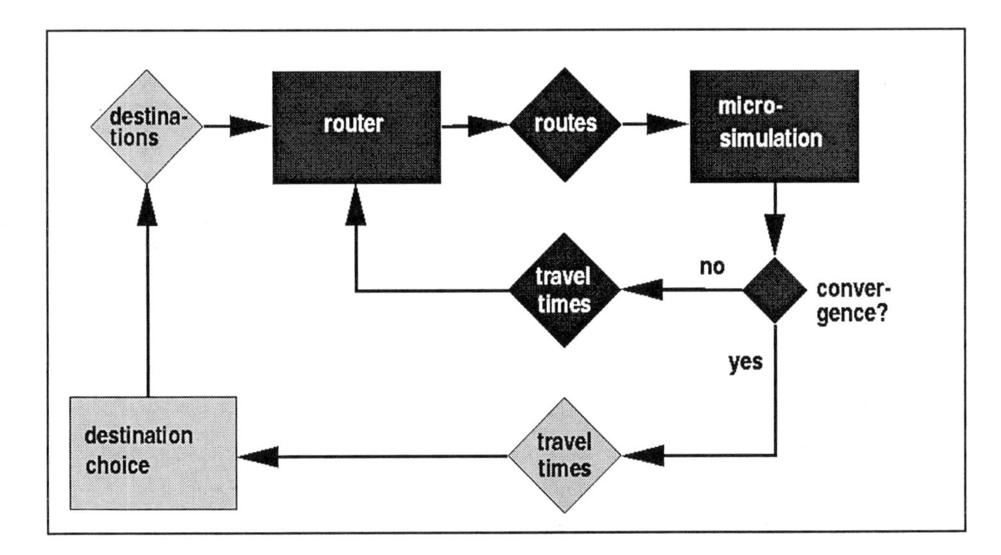

Figure 1: Scheme of the models and data needed to compute a transportation system.

Currently, the FVU uses a simplified version of this approach where an OD–matrix is determined for every hour by the classical approach. This OD–matrices are then decomposed into individual trips which serve as an input for the combined route choice and travel simulation. The computation of the OD–matrices is described in detail in the article by K.J. Beckmann. The rest of this treatise only deals with the route choice and the traffic flow simulation (darker shade in Fig. 1). Moreover, we will neglect all other travel modes such as public transport or bicycle traffic completely. They are an important part of the modelling of human activities as demonstrated in the articles by Doherty and Timmermans in this book. They can be included in the route-choice and simulation process as well by using the formal language approach to route-finding [4] currently implemented into TRANSIMS.

2 Simulation–Based Route Choice

Finding the routes through a given network with time-dependent demand, the so-called dynamic traffic assignment problem (DTA), can be mathematically formulated as an optimisation problem with linear constraints [5]. Using the assumption that individual travellers try to find their shortest or cheapest route through a network [6], the problem is complicated because the travel times on a link depend on the number of cars on the link. When dealing with a dynamic problem, the classic approach using time-independent flows is no longer suitable, because a number of additional dynamic constraints have to be taken into account. To name only the most important ones: (1) spill-back phenomena have to be described and (2) the so-called FIFO-condition (First-In, First-Out) has to be fulfilled. FIFO means that a car which enters a link first will also have to leave this link first. Within the framework of a description by flows this is difficult to model; when using a microscopic approach with individual travellers, this becomes trivial. To compute a static traffic assignment, an iteration process is needed that successively adds new routes to any of the OD–relations within the network. The dynamic microscopic approach needs a similar iteration. In each iteration step a complete simulation of the transportation system over a certain time-period is performed, leading to the travel times needed by the individual travellers. (Therefore the name simulation-based DTA.) Any traveller has either a set of routes which are tested in subsequent iteration steps until she is satisfied [7], or in any iteration step some of the travellers acquire new routes based on the travel times of the iteration step before [8]. In the FVU we pursue the first approach whose underlying strategy will be described in greater detail later on in this article.

This simulation-based DTA can be used to solve static problems as well. The idea of performing a full microscopic simulation in each iteration step seems rather wasteful. However, preliminary results suggest [9] that it can be done nearly as fast as the classic stochastic static traffic assignment. This is the case if the static problem is solved with full mathematical rigor and none of the somewhat hard to justify approximations often applied to practical algorithms are used. For the dynamic problem up to now no implementation is known to us that can solve larger networks than some 300 nodes exactly. The simulation-based DTA, however, can tackle up to 10^5 links with 10^7 cars in a reasonable time on a parallel computer (Of course, this depends on the microscopic model used.).

The DTA algorithm used works as follows. Any traveller has a (usually small) set of routes to choice from. Associated with the routes is a probability to choose this route. After assigning a route to any of the travellers according to these probabilities, a simulation is carried out that leads to the actual travel times. These actual travel times are used to shift the probabilities towards shorter routes. Convergence is reached, when these probabilities didn't change any more. The exact formulation of the shift in the probabilities can be found in [7]. It is interesting to point out that this approach can be translated completely into the parlance of game theory: the routes correspond to the strategies in a dynamic N–person game where each agent follows a so called mixed strategy approach. Using the approach, where each traveller has only one route is a dynamic N–person game with pure strategies [10].

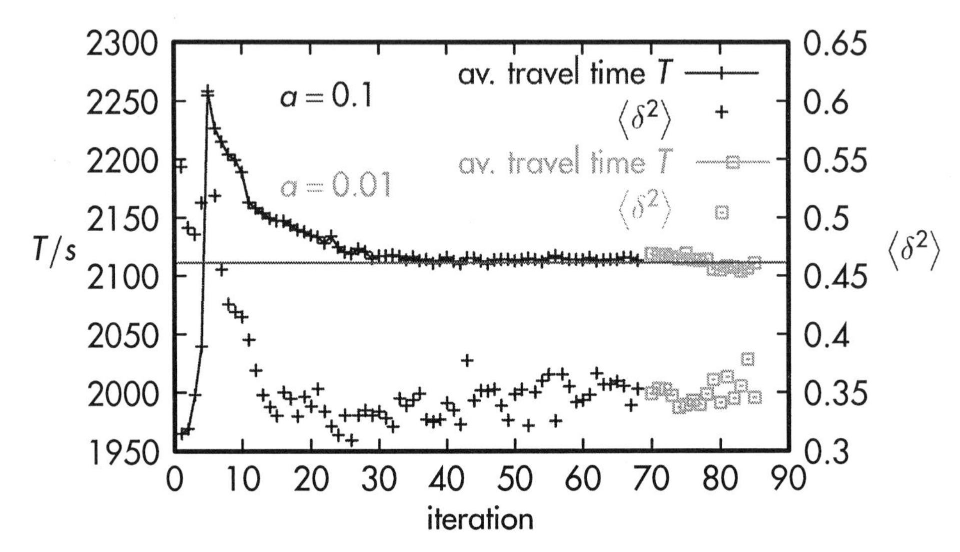

Figure 2: Convergence of a simulation-based DTA for the road network of Wuppertal.

3 Simplified Modelling of Traffic Flow

The realistic model of traffic flow to be described in the next section has one shortcoming: although it is very fast, for the repeated simulation to solve the DTA-problem, it is still faster simulation would be helpful. Therefore, for doing an initial assignment of routes, another model is used that is even faster than the realistic model. To have a microscopic and dynamic description of traffic in a network, it is sufficient to model FIFO, spill-back, finite capacities and finite travel-times on a link. One of the simplest models that could handle this is one based on the concept of queues. Each link of the network is modelled as a waiting queue with exit constraints and a certain travel time which may depend on the number of cars already on the link. Additionally, if a car tries to enter a congested link, it is denied access, therefore spill-back phenomena are very easily included in the description. This model has two advantages, and one disadvantage:

1. It is very fast, for certain applications two orders of magnitude faster than the model in the subsequent section.
2. On the macroscopic level of flow through a link it can be calibrated to mimic any more realistic microscopic model (see Fig. below).
3. However, not all of the phenomena of daily traffic are captured within this model, and it is difficult to compute emissions from its simplified dynamics.

Nevertheless, it can be used to determine a first set of routes, so that only a few additional iterations with a realistic microscopic model are needed.

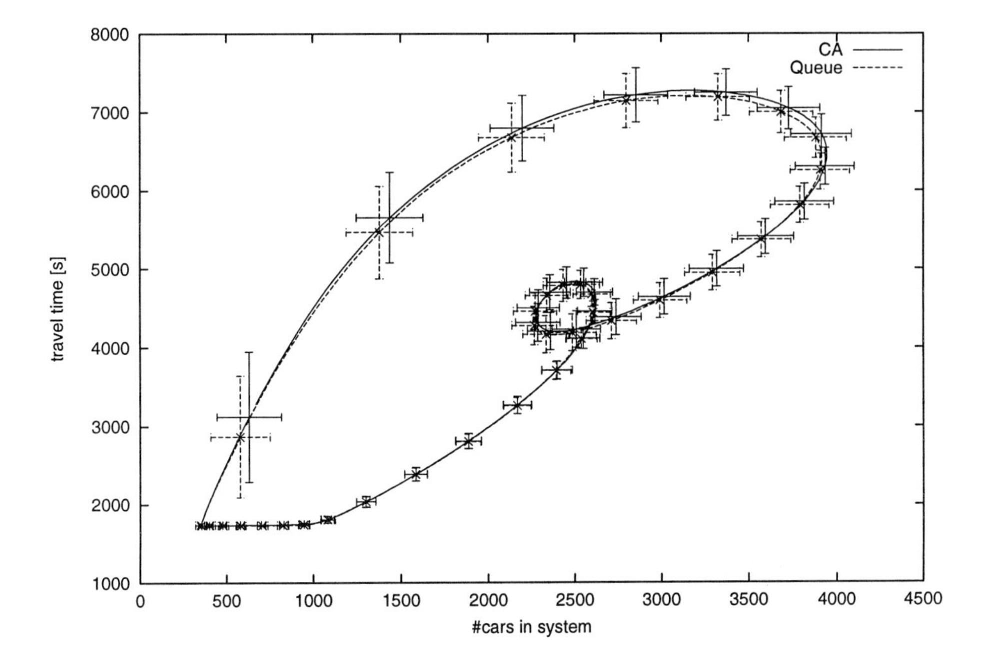

Figure 3: Travel times resulting from micro simulation and queuing model after calibration.

4 Realistic Modelling of Traffic Flow

In order to compute the emissions caused by traffic, and to take care of a number of phenomena not described by the queue-model of the last section, a more detailed account on the microscopic aspects of car-following is needed. There is an enormous multitude of models on the market, most of them only poorly understood in the important aspect of which microscopic feature implies which macroscopic behaviour [11–13]. Since the details of human driving behaviour are intricate, it is good modelling strategy to avoid going into too much details. The model described in the following models very few features of car-following:

1. Finite maximal velocity v_{max},
2. Finite acceleration a bounded by $a \in [-b_{max}, a_{max}]$,
3. Crash–freeness.

As a side-remark is has to be mentioned that the restriction to a maximum deceleration is not absolutely necessary. Traffic flow can be described without this, as can be seen in the article by M. Schreckenberg. The three presuppositions above, together with the time discrete update of the dynamic state of the cars (time-step t, position $x(t)$, velocity $v(t)$ and lane $l(t)$) lead to the following update formula, which follows from a simple consideration about the braking distances:

$$v_{Des} = \min\{v(t) + a_{max}, v_{Safe}, v_{max}\}, \quad (1)$$

$$v(t+1) = \max\{0, v_{Des} - \xi\}, \quad (2)$$

$$x(t+1) = x(t) + v(t+1), \quad (3)$$

where ξ is a noise term modelling the tendency of real drivers not always driving optimally and v_{Safe} is computed according to:

$$v_{Safe} = v_{Leader} + (g(t) - v_{Leader}) / (\tau + 1), \quad (4)$$

where v_{Leader} denotes the velocity of the car ahead, $g(t)$ the headway, i.e. the distance between the two cars, and τ is a „relaxation“ time given by

$$\tau = (v(t) + v_{Leader}(t)) / (2\, b_{max}) . \quad (5)$$

The lane-changing logic, which especially for German freeways has to take into account the so-called lane-inversion phenomenon (at high traffic densities the left lane is used more heavily than the right lane), is based on two rules [14, 15, 11]:

1. If a car is hindered by the car ahead, change to the left.
2. If there is plenty of space in front on the right and left lane, go back right.

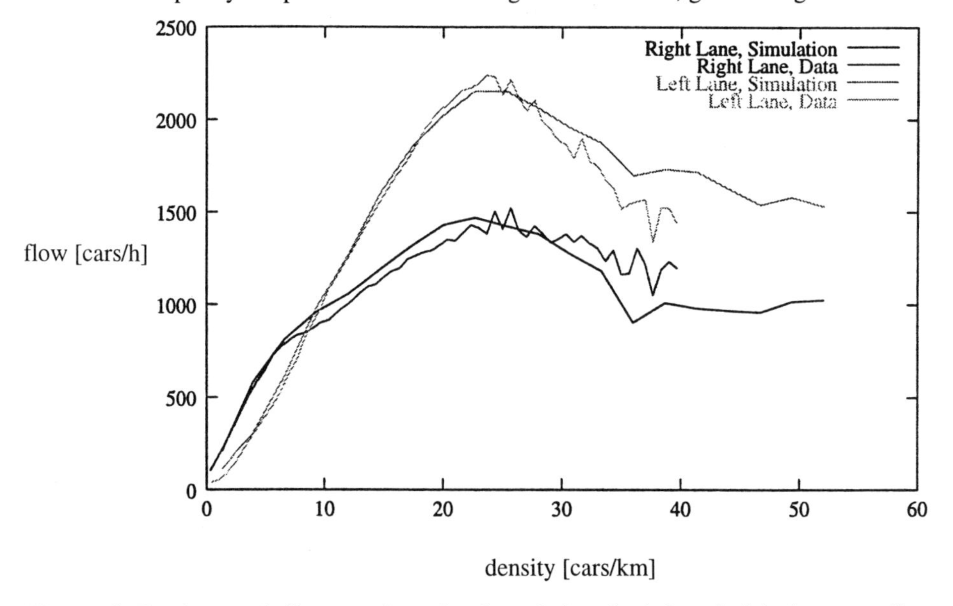

Figure 4: Fundamental diagram, flow-density relation for left and right lanes on German highway, comparison between data and simulation.

A car is allowed to change only if the system remains collision-free after the lane-change. This simple model work very well for freeway traffic, as can be seen in Figs. 4 and 5 respectively. Fig. 4 shows the fundamental diagram of the model, together with data. The agreement is very well. Fig. 5 shows a dynamic situation: as measured by the three counting devices, a jam is travelling backward through the freeway segment under consideration. Two of the data-sets measured by the counting devices were used to drive the simulation, while the data-set of the middle counting device served to compare the simulation to reality. Although the simulation didn't catch any of the wiggles of the measured data, the agreement is reasonably good. Additional calibration (results not shown here) lead to the conclusion that this model can be used with confidence in order to compute the emissions caused by traffic.

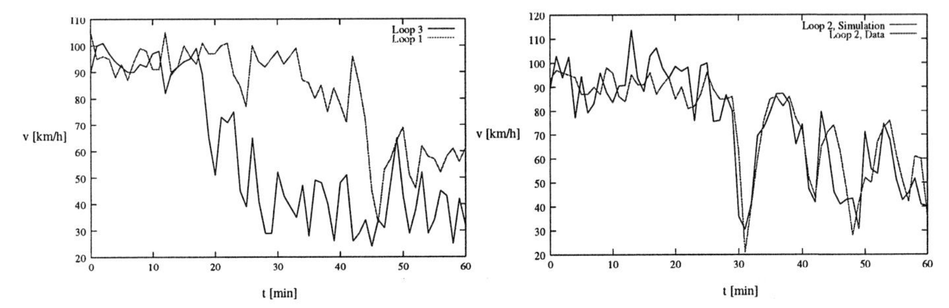

Figure 5: Reconstruction of loop data: On a highway segment (3km) data from loops 1 and 3 are used as input for the micro simulation. The second plot shows a comparison between data and simulation for the speed of cars passing the middle loop.

5 Projects

A number of projects have been done with the tools at hand, some of them yet to be completed:

Fig. 6 shows the result of a simulation of the German freeway network, driven by a static OD–matrix provided by the DLR. Unfortunately, this OD–matrix does not include transit traffic, so border-crossing traffic volumes are too low and a comparison with real world data is difficult. Nevertheless, the overall traffic pattern looks plausible.

Simulation of the freeway network of Germany with a static OD–matrix.

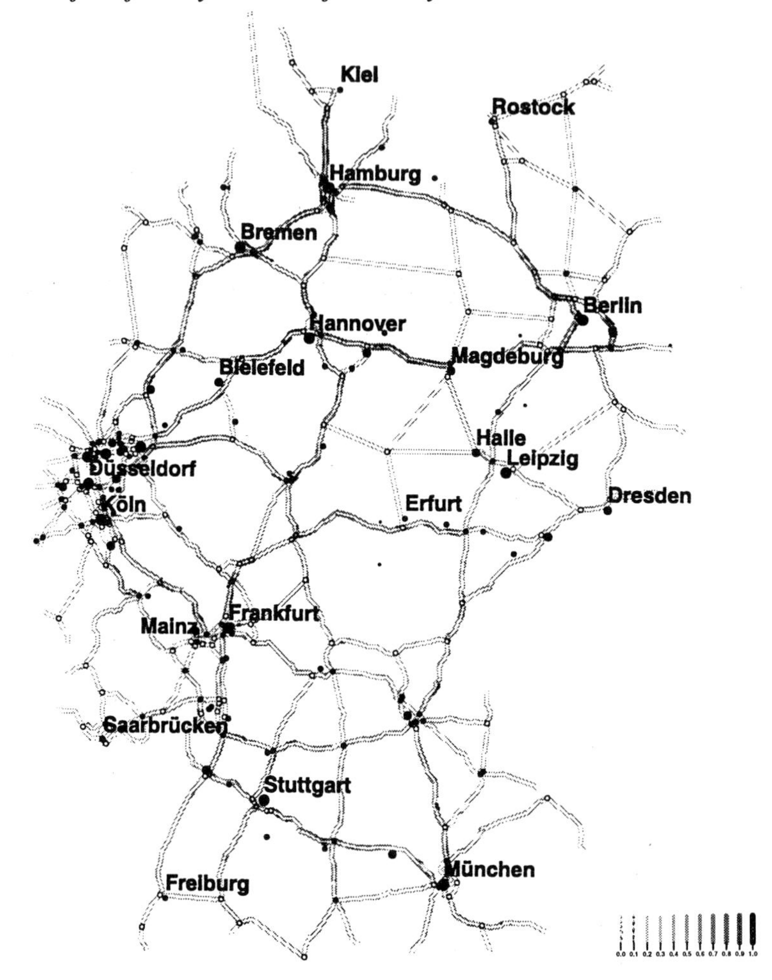

Figure 6: Results of a simulation of the German freeway network with a static OD–matrix provided by the DLR. Note that transit traffic is missing in this OD–matrix, resulting in unrealistic low traffic volumes near the border.

Fig. 7 compares the results of a static and dynamic assignment for a small urban road network for moderate traffic demand. It turns out that the differences in this case are rather small. Differences between static and dynamic assignment only show up when the demand exceeds the capacity during the rush-hour and the link travel times cannot be adequately described as a static function of the link flows.

Comparison between the DTA and the classic four-step-approach for a small town

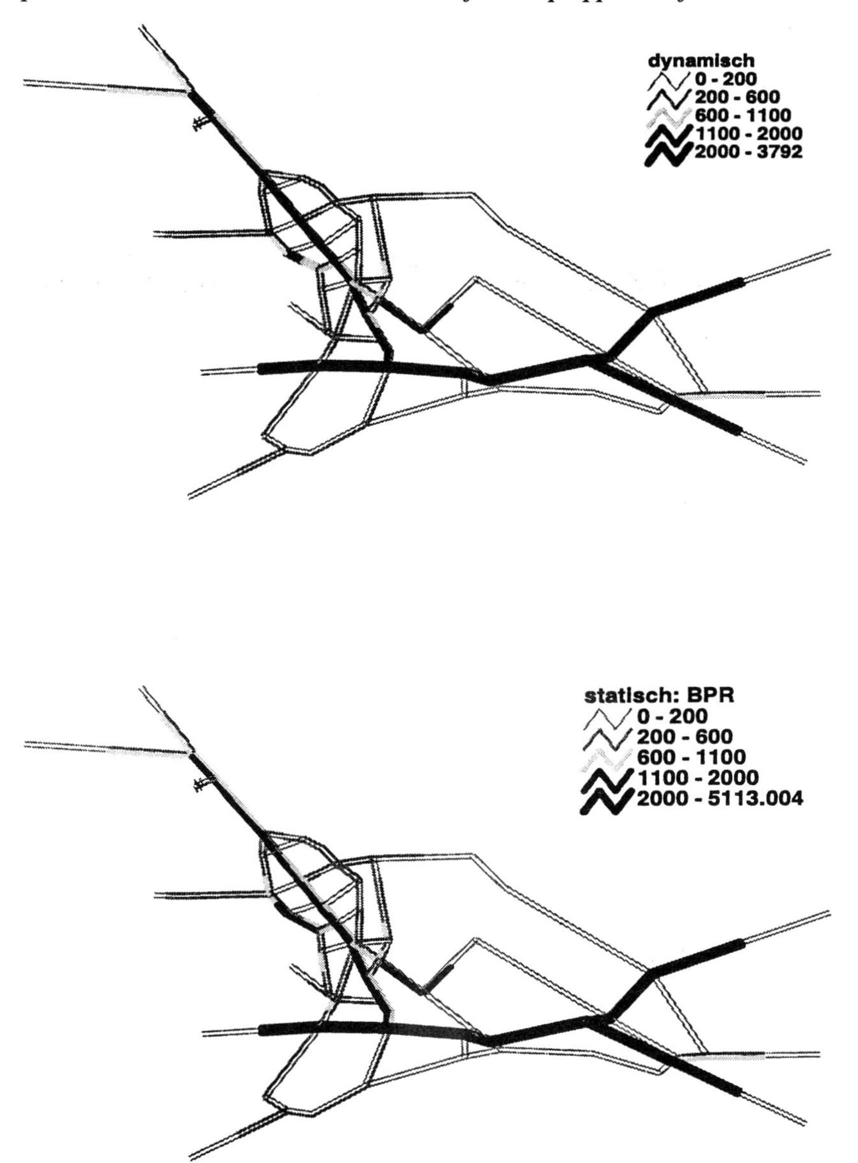

Figure 7: Comparison between static and dynamic assignment for a small urban network. For moderate congestion levels, the differences are negligible.

Simulation of the freeways of Nordrhein–Westfalen with a dynamic OD–matrix

In Fig. 8, the result of a simulation based assignment is shown for the highway network of Nordrhein-Westfalen. The dynamic OD–matrix was provided by the ISB.

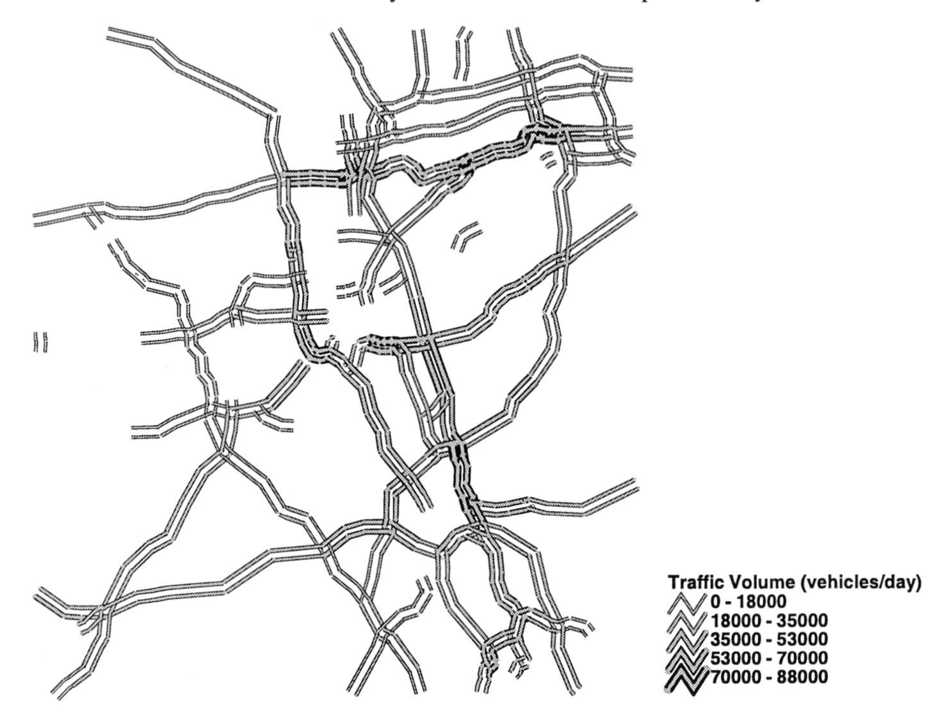

Figure 8: Results of a dynamic, simulation based assignment for the freeway network of Nordrhein-Westfalen. Shown are the average daily traffic volumes.

The differences between simulated traffic volumes and measured data are within 10%.

Simulation of a medium-sized city (Wuppertal, 400,000 inhabitants) with a dynamic OD–matrix.

Fig. 9 shows a screen-shot of a micro-simulation of the network of Wuppertal, driven by a dynamic OD-matrix. Even though the network has over 9000 nodes and over 16000 links, the simulation runs in real-time on a 8 processor Sun Enterprise 4000.

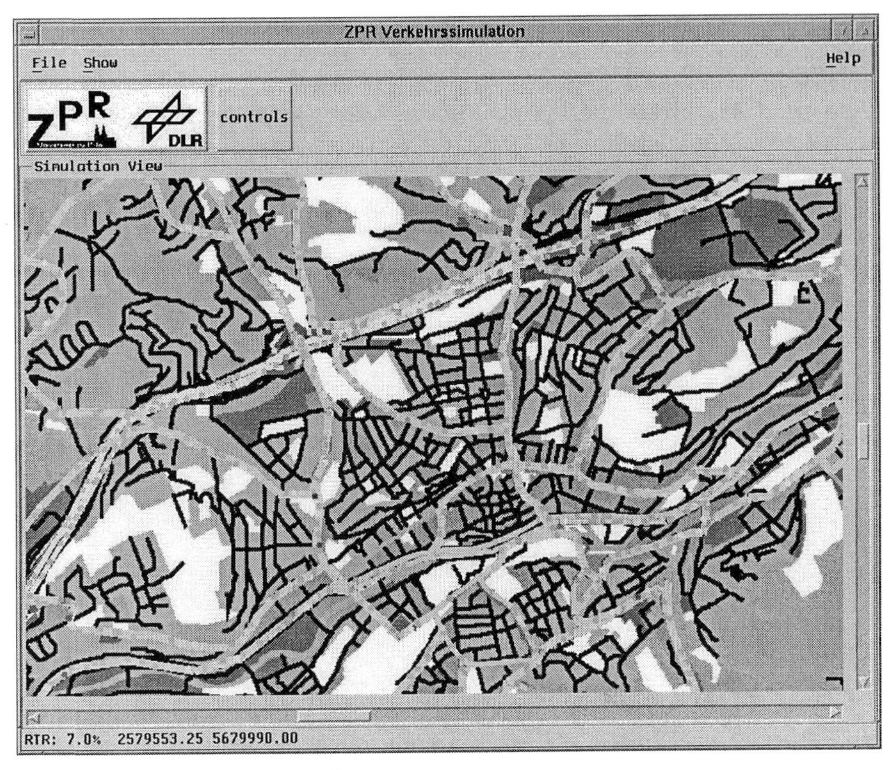

Figure 9: Micro-simulation of Wuppertal. (Screen-shot, showing individual vehicles).

6 Destination Choice – a Challenge for Microscopic Modelling

Since the classical means for computing time-dependent OD–matrices are not very well suited for the purpose of the micro-simulation of transport systems, first attempts have been made in modelling the demand for transport on the basis of single individuals. The basic idea is very simple: try to build the daily activity chain of any individual that lives within the simulation area. This can be done on the basis of socio-demographic household data that are available for at least for the US–society on a very fine-grained scale. Doing so requires some understanding about human decision making, which again is a topic related to game theory. However, other modelling approaches to this problem exist and are currently the topic of intense research, as can be seen in the contribution of Timmermans in this volume. Currently, an attempt is made to implement the classic modelling on the basis of individuals, which already removes some of the weaknesses of the classic approach. Results on this topic will be presented elsewhere.

First results that have been obtained in co-operation with the TRANSIMS project which will be summarised in the following [18]. There the problem of assigning employees to their respective working places have been tackled with a gravity-type approach. The city is divided into small cells of blocks each containing some hundred inhabitants. The inhabitants in each block are generated by an algorithm that uses socio-demographic information to generate a synthetic population with the same statistical features as the original one. Each block has therefore a number of employees, and, from another data-source, the number of working places. Information about where the employees are working is not available and has to be generated by a model. The algorithm assigns any employee to one of the working places probabilistically, where the probability of assigning an employee to a certain block is proportional to the number of free working places time a factor $Q(\tau)$ that depends on the travel-time τ to this place. After having assigned one of the employees to a working place, it is labelled not free and no longer used in the assignment process.

The great advantage of this microscopic approach lies in the fact, that it is known exactly which individual go to which working place, and that (s)he can be send back to his/her living place (which is not the case for the classic four-step-algorithm). Furthermore, this approach generates exactly the input needed for a microscopic simulation. This simulation can then be used in order to give further constraints on the destination choice process, since it erases the trips which are impossible. This elimination is done by a re-assigning process. This re-assignment either eliminates trips that are too long, or it prunes a number of routes that together generate too much traffic in the city.

To judge the quality of this process, the probability distribution of travel times has been used. For the city of Portland, for which this work has been performed, information about the true travel times is available. After several iterations done with the combined destination/route choice process, the results in Fig. 10 were achieved. It can be seen that the process slowly approaches the empirical distribution, however up to now it didn’t converge completely, indicating that further research has to be done to speed up convergence.

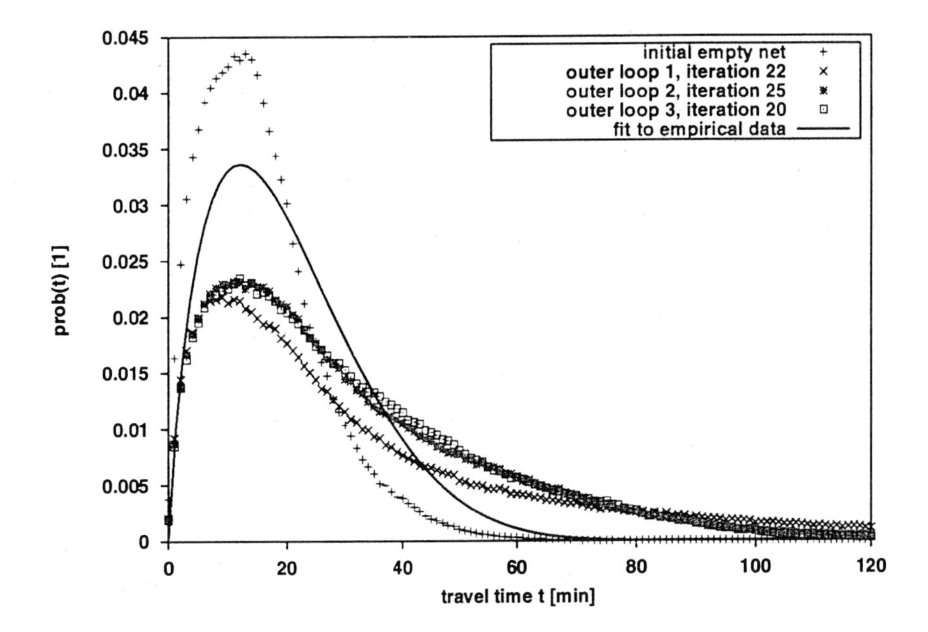

Figure 10: Travel time distributions for the empty network, and a number of intermediate iteration steps in the combined destination/route-choice process, compared to the empirically measured one.

7 Summary

The dynamic and microscopic approach described in this article is definitely ready to find a practical application, at least for the part of the route choice and the travel modelling. It remains to be done to construct similarly efficient tools for the activity description as well. Although promising approaches exists, there is still a lot of work to be done in this field. What can be gained from succeeding in this efforts is an integrated dynamic simulation of the transport system of a large city. A further benefit can be obtained from learning how to deal with the modelling and simulation of large socio-economic systems. Sooner or later it may happen that the simulation of the transport system as sketched here and as is done in the FVU and at other places is the part of a still larger simulation effort [19] which tries to describe any economic activity in a city, therefore giving the business of the planning of cities a completely new basis.

Acknowledgements

A number of people have contributed to the results presented here. We would like to thank A. Bachem, S. Hasselberg, S. Janz, Ch. Moll, P. Oertel, A. Schadschneider, A. Stebens, and A. Vildósola for their help and for their readiness for discussions. We thank all members of the FVU who helped to understand the subject presented here by numerous discussions. The FVU has been a big adventure in interdisciplinary work, at least for us.

References

1. FVU–homepage, http://www.zpr.uni-koeln.de/Forschungsverbund-Verkehr-NRW/.

2. TRANSIMS–homepage, http://studguppy.tsasa.lanl.gov/.

3. SMARTEST–homepage, http://www.ist.leed.ac.uk/smartest/; see also [1].

4. Ch. Barrett, R. Jacob, and M. Marathe, Formal Language Constrained Path Problems, SWAT (1998), preprint can be obtained from [2].

5. B. Ran and D. Boyce, *Modeling Dynamic Transportation Networks*, Springer, 2nd edition, 1996.

6. J. G. Wardrop, Some Theoretical Aspects of Road Traffic Research, Proceedings of the Institute of Civil Engineers, vol. 1 (1952).

7. Ch. Gawron, Int. J. Mod. Phys. C **9**, 393 (1998).

8. M. Rickert, Ph.D.-thesis, can be obtained from ZPR–homepage http://www.zpr.uni-koeln.de/ (1998).

9. A. Vildósola, private communication, and Diploma-thesis (1998).

10. J. Wiebull, *Evolutionary Game Theory*, MIT Press, 1995.

11. S. Krauß, *Microscopic Modeling of Traffic Flow: Investigation of Collision Free Vehicle Dynamics*, Ph.D.–thesis (1998).

12. S. Krauß, P. Wagner, and Ch. Gawron, Phys. Rev. E **54**, 3707 (1996).

13. S. Krauß, P. Wagner, and Ch. Gawron, Phys. Rev. E **55**, 5597 (1997).

14. P. Wagner, K. Nagel, and D.W. Wolf, Physica **A 234**, 687 (1996).

15. K. Nagel, D.E. Wolf, P. Wagner, P. Simon, "Two-lane traffic rules for cellular automata: Asystematic approache", Phys. Rev. E **58**, 1425-1437.

16. Yosef Sheffi, *Urban Transportation Networks*, Prentice Hall, Englewood Cliffs 1985.

17. D.F. Ettema & H.J.P. Timmermans (eds.), *Activity-based Approaches to Travel Analysis*, Pergamon, 1997.

18. P. Wagner and K. Nagel, submitted to Transportation Research Board 1999.

19. E. J. Miller and P.A. Salvani, ILUTE, in 77. Transportation Research Board 1998.

Evaluation of Cellular Automata for Traffic Flow Simulation on Freeway and Urban Streets

W. Brilon and N. Wu

Lehrstuhl für Verkehrswesen, Ruhr-Universität Bochum, 44780 Bochum, Germany

A Cellular Automaton is a extremely simplified program for the simulation of complex transportation systems, where the performance velocity is more important than the detailed model accuracy. The first application of the Cellular Automaton for simulation of traffic flows on streets and highways was introduced by Nagel and Schreckenberg [7]. The basic Cellular Automaton model from Nagel-Schreckenberg has been checked against measurements of realistic traffic flow on urban streets and motorways in Germany. It was found that the measured capacities on German motorways cannot be reproduced very well. On urban streets, however, it was very well possible to represent traffic patterns at intersections. The paper describes a completely new concept for the cellular automaton principle to model highway traffic flow. This model uses a time-oriented car-following model. This model accounts for the real driving behavior more precisely than the model from Nagel and Schreckenberg. This paper shows that a Cellular Automaton is generally applicable for simulation of traffic flows. The degree of correspondence with reality depends on the applied car-following model. The new model concept combines realistic modeling with fast computational performance.

1 Introduction

Over the past decades, the total demand for transportation in the developed countries is increasing, especially within the highway networks. Traditionally the supply with more and wider motorways helped to cope with increasing demand. This option, however, is not longer available due to environmental and budget restrictions. Thus, more intelligent methods to enhance the capabilities of highways are desired. Computer simulations may help in the task of planning more complex transport systems.

A very fast simulation model is the so-called Cellular Automaton (CA), which can be used for large-scale networks and because of the immense speed capability of the model even for traffic assignment and traffic forecasting purposes. For instance, in Germany and in the USA attempts are made to use vehicle by vehicle assignment by CA to predict the traffic congestion in urban and motorway networks.

The objective of the CA was set to running as fast as possible. Therefore, the so-called standard CA-model (STCA, or the Nagel-Schreckenberg [7] model) uses only four simple rules for the car-following behavior and has only one system parameter to be calibrated. The car-following model in STCA is space-oriented and is more or less of heuristic nature. Surprisingly, the results of the STCA for queuing systems agree very well with the theoretical approaches despite of the simple rules. Therefore, the traffic behavior at intersections on urban streets which are commonly considered as queuing systems can be reproduced by the STCA satisfactorily. Because of the simplification of the car-following model in STCA, the traffic flow on motorways, especially on German motorways, cannot be described with sufficient performance by the STCA. Observed speed-flow relationships on German motorways cannot be reproduced very well.

In this paper, some major results for the calibration and validation of the STCA is presented. The parameters of the STCA are, as good as possible, given the different types of road and intersection.

In addition, a new CA-model (TOCA), which uses a time-oriented car-following model, is introduced. This new model accounts for the driver behavior more realistically than the space-oriented STCA does. It retains all of the positive features of a CA and modifies only the car-following behavior of drivers. The TOCA is verified with data measured on German motorways. Realistic capacity and speed-flow relationships can be reproduced with better performance.

2 The Standard Cellular Automaton (STCA)

2.1 The Model and its Rules

Designing a simulation model as simple as possible, the most radical way is to use integer variables for space, time and speed. Such a simulation model is called a cellular automaton [14]. The space is divided into cells that can contain a vehicle or can be empty. The length of a cell is given by the minimum space headway between vehicles in jam. It is the inverse value of the jam density k_j and is set to be 7.5 m ($k_j = 133$ veh/km). The update time-step Δt is rather arbitrary, one usually uses the driver reaction time that lies between 0.6 and 1.2 s. Thus, also Δt can be considered as a parameter to be calibrated. The speed ranges from 0 to $v_{max} = 6$ cells/Δt. That corresponds to a speed of 162 km/h for $\Delta t = 1$ s. The STCA uses a simple philosophy that approximately describes the dynamic of driving (car-following): go as fast as you wish and as the vehicle in front allows you and decelerate if you have to avoid a rear end collision [7, 6].

This philosophy is represented by the following four rules in the STCA. Denoting *gap* as the number of free sites in front of a vehicle, *x* the actual position and *v* the actual speed of the vehicle, one obtains (cf. Figure 1) the pseudo-code for the rules:

(1) If ($v < v_{max}$) then $v = v + 1$ (if the present speed is smaller than the desired maximum speed, the vehicle is accelerated). The desired speed x_{max} can be assumed to be distributed by a statistical distribution function where the values of v_{max} are only allowed to be 1, 2,.., 6 cell/Δt.

(2) If (v > gap) then v = gap (if the present speed is larger than the gap in front, set v = gap). This rule avoids rear end collisions between vehicles. Note that here a very unrealistic braking rule allowing for arbitrarily large decelerations is involved. This rule forces a minimum time headway of Δt s.

(3) If (v > 0) then v = v – 1 with p_{brake} (the present speed is reduced by 1 with the probability p_{brake}). This rule introduces a random element into the model. This randomness models the uncertainties of driver behavior, such as acceleration noise, inability to hold a fixed distance to the vehicle ahead, fluctuations in maximal speed, and assigns different acceleration values to different vehicles. This rule has no theoretical background and is introduced quite heuristically. The most of the shortcomings of the STCA are due to the unrealistic rule.

(4) x = x + v (the present position on the road is moved forward by v)

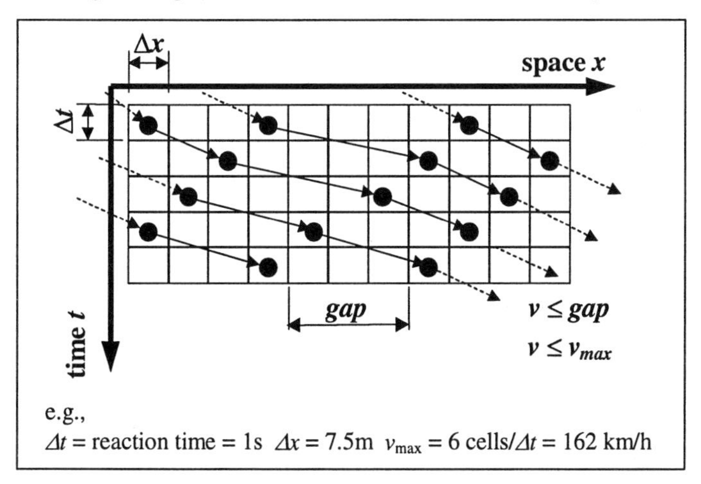

Figure 1: Principle of a Cellular Automaton.

According to theses rules the speed and the acceleration/deceleration ratio of a vehicle are independent of speed of other vehicles at any time. They are only functions of the gap in front. Thus, these rules can be updated in parallel for any vehicle. However, the acceleration and deceleration ratio can take infinite large value if a vehicle changes its speed according to these rules. The average deceleration ratio over the driver population is p_{brake}. The average acceleration ratio over the driver population yields $1\text{-}p_{brake}$. Despite its extreme simplicity, this model shows many features which agree with the real-world traffic. Especially, the distribution of time headways and the distribution of arrivals within a time interval can be somewhat satisfactorily reproduced. The STCA of Nagel-Schreckenberg uses a minimal set of rules that under certain conditions yield desired macroscopic behavior. For describing queuing systems, e.g., intersections of two urban streets, the STCA delivers very good results compared with the real-world traffic conditions [2].

2.2 Application of STCA for Urban Streets

For streets in urban traffic networks, intersections (with or without traffic signals) are always the bottlenecks of the total system. The capacity of the intersections are therefore decisive for the total network.

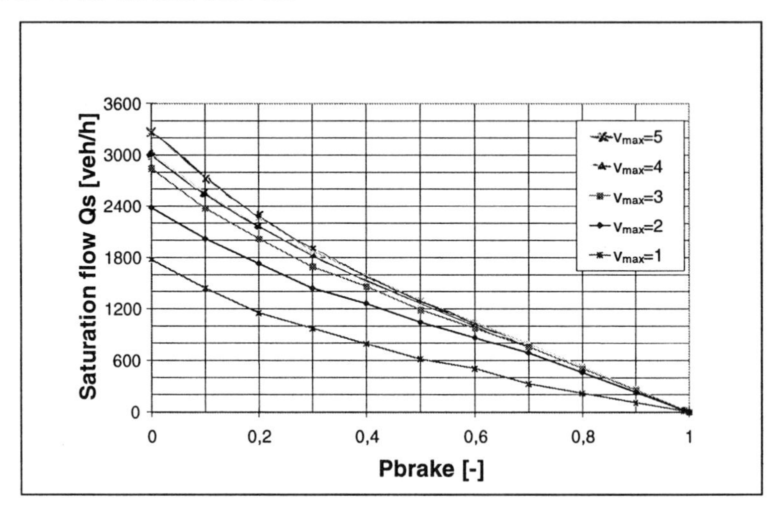

Figure 2: Saturation flow Q_s from STCA at signalized intersections.

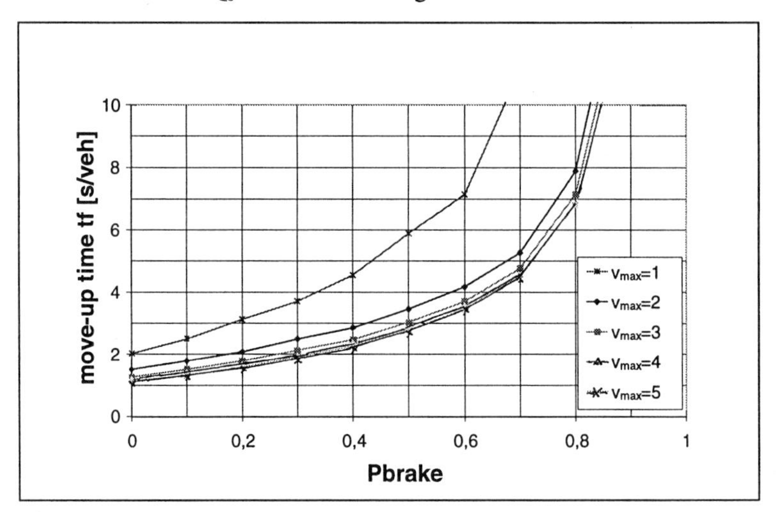

Figure 3: Move-up times t_f from STCA at unsignalized intersections.

At intersections with traffic signals, the capacity of the intersections can be reproduced very well with STCA by varying the parameter p_{brake}. Figure 2 shows the shape of the capacity curves at intersections with traffic signals as function of the parameter p_{brake}. The curves are organized according to the desired speed (v_{max}) at the intersection. These curves indicate the simulated discharge capacity of a queue by STCA. With this curves, the saturation floe during green time (Q_s) at an intersection with traffic signals can be easily reproduced by choosing a occasional parameter p_{brake}. For instance, by using a $p_{brake} = 0.2$, a saturation flow of about $Q_s = 1700$ veh/h can be obtained for $v_{max} = 2$ cell/Δt (corresponds to V = 54 km/h). In other words, if a saturation flow of $Q_s = 1700$ veh/h for $v_{max}=2$ cell/Δt is needed, a parameter $p_{brake} = 0.2$ should be applied in STCA.

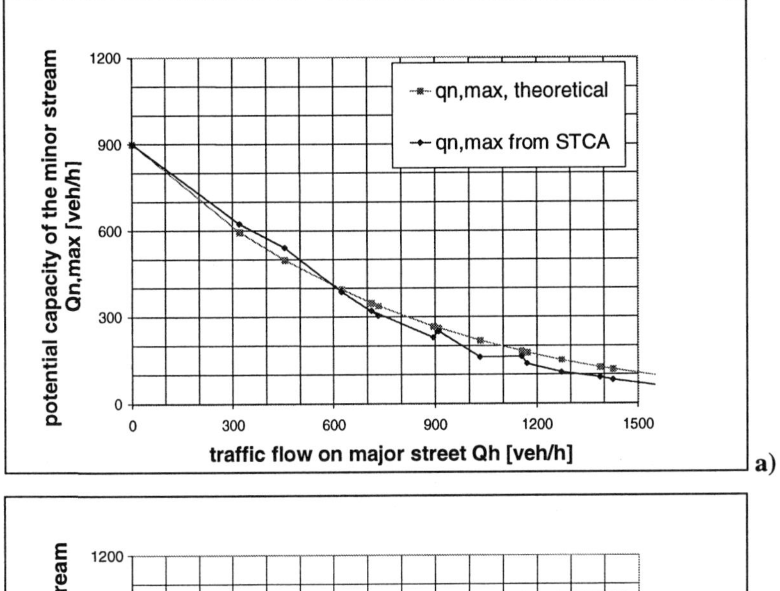

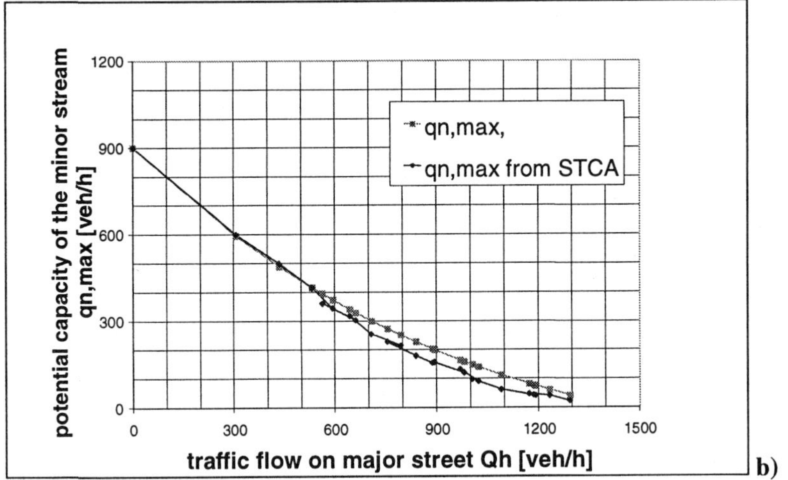

Figure 4: Potential capacities from STCA at unsignalized intersection
a) potential capacity for single-lane major-streets
b) potential capacity for multi-lane major-streets.

From these curves of capacities, also the move-up times t_f at intersections without traffic signals (i.e., priority-controlled intersections) can be calculated. The move-up time t_f represents the average time headway between two vehicles departing at the stop line in succession. It is the reverse value of the capacity at the stop line with zero major street flow. Therefore, one obtains $t_f=1/Q_s$.

According to this relationship the curves for the move-up times t_f can also be obtained from the Figure 2. The curves for t_f are directly shown in Figure 3. The move-up times t_f are functions only of the parameter p_{brake} for the subject minor street.

The critical gaps t_g at intersections without traffic signals can be considered as a function of the number T of cells which have to be checked by the vehicle waiting at the intersection, the desired speed $v_{h,max}$ and the parameter $p_{brake,h}$ for the major street. The following relationship can be stated: $T = t_g \cdot (v_{h,max} - p_{brake,h}) - 1$. For instance, if a critical gap $t_g = 5$ s, a speed on the major street $v_{h,max} = 2$ cells/Δt (corresponds to $V = 54$ km/h), and a parameter $p_{brake,h} = 0.2$ for the major street are used, one has to apply $T = 5 \cdot (2 - 0.2) - 1 = 8$ cells in STCA.

Thus, according to the procedure: a) setting the parameter $p_{brake} = f(t_f)$ for the minor street (cf. Figure 3), b) setting the parameter $T = t_g \cdot (v_{h,max} - p_{brake,h}) - 1$ for the major street, and c) simulating the capacity $Q_{n,max} = f(t_f, t_g)$ using p_{brake} and T, the potential capacity of the minor street at intersections without traffic signals can be obtained by STCA.

The potential capacity simulated with this approach by STCA at intersections without traffic signals agrees very well with the theoretical results (cf. Figure 4) obtained from gap acceptance theory.

Therefore, the using of this approach STCA can be recommended for simulating large scale urban street networks. Here the intersections always represent the bottlenecks of the system whereas the traffic flow on links is nor decisive. Thus the simple model is sufficient for moving the vehicles along the links of the networks to represent real world conditions.

2.3 Application of STCA for Freeways

On the other hand, as a result of the extreme simple configuration of the STCA, the characteristics of traffic flow in motion (no queuing) can cannot be described by STCA with satisfying performance. The simulated traffic flow on motorways, especially on European motorways, cannot represent the realistic speed-flow relationships. Analyzing the STCA in details one finds the following unrealistic features on the microscopic level:

1. The acceleration and deceleration ratios are unrealistic because of their infinite value and dependence on each other (acceleration ratio $=1-p_{brake}$, deceleration ratio $=p_{brake}$).
2. The acceleration ratio ($=1-p_{brake}$) is larger than the deceleration ratio ($=p_{brake}$) (for $p_{brake} > 0.5$).
3. The speed of a vehicle is not dependent on the speed of other vehicles ahead.

4. The minimum time headway between two vehicles cannot obtain values smaller than Δt.

5. The driver reaction time is always equal to Δt.

6. The speeds are classified in discrete classes (i.e. with steps 27 km/h).

7. The threshold (time interval to the vehicle ahead) for changing (adjusting) speed is always equal to Δt.

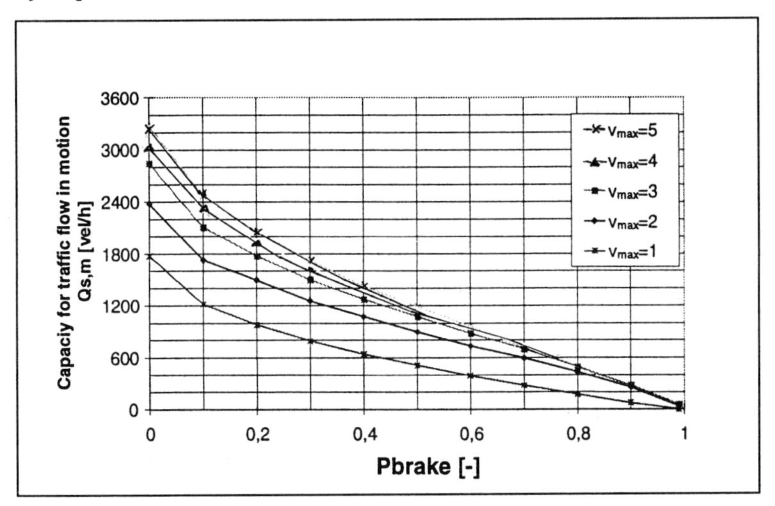

Figure 5: Capacity $Q_{s,m}$ from STCA for traffic flow in motion.

These microscopic properties of the STCA lead to the following macroscopic disagreements compared to real traffic flow on motorways:

1. The maximum flow ratio in the opening (depressive) phase C_{open} is larger than the maximum flow ratio in the closing (compressive) phase C_{close}, whereas in reality just the opposite occurs.

2. No capacity drops exist.

3. The convoy dynamics is always stable.

4. No breakdowns in an open pipeline system.

Figure 5 shows the curves of capacities $Q_{s,m}$ simulated by STCA for traffic flow in motion. Comparing the queue discharge capacities (saturation flow Q_s) for queuing systems (Figure 2), it can be recognized that the capacities for traffic flow in motions are always smaller than the queue discharge capacities. This is the main reason why in the car-following model of STCA the maximum flow ratio in the opening phase C_{open} is larger than the maximum flow ratio in the closing phase C_{close} and why no capacity drops and breakdowns occur.

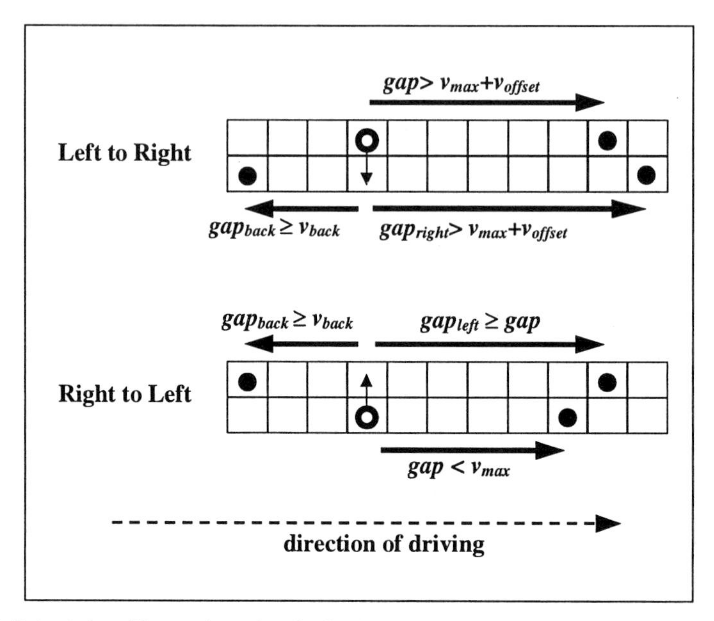

Figure 6: Principle of lane-changing in STCA.

For simulation traffic flow on freeways, configurations with more than one lane should be taken into account. In order to simulate multi-lane traffic an additional set of rules for lane-changing is needed. These rules are updated before the speed update is done. They are divided into two parts representing the desire and the possibility of changing lane without hindering the other vehicles. Also, the reason of changing from left to right is different from the reason of changing from right to left [11, 12, 8]. Because of the so-called "mandatory driving on the right lane and the prohibition of overtaking on the right" (MDR) on motorways in most of the European countries the lane-usage between different lanes is non-homogeneous (on US motorways the lanes are used more or less homogeneously under heavy traffic volume due to the rule "keep in lane"). The pseudo-code of these rules in STCA for changing lane in case of MDR are (cf. Figure 6):

(1) if (v_{max} > gap) and (gap_{left} >= gap) then desire-from-right-to-left

(2) if (v_{max} < gap - v_{offset}) and (v_{max} < gap_{right} - v_{offset}) then desire-from-left-to-right

(3) if v_{back} <= gap_{back} then possibility-of-changing-lane

Here, v_{offset} is a parameter which is used to represent the asymmetric driving and distance behavior on different traffic lanes. This parameter then significantly affects lane split of traffic flow on motorways. In addition, a ban on passing on right is implemented. This can be written as

(4) if (v_{right} > gap_{left}) then v_{right} = gap_{left} (prohibition of passing on right)

3 The New Time-Oriented Cellular Automaton (TOCA)

The major problem of the STCA is the unrealistic driver behavior caused by an unrealistic car-following model. Considering the rules of STCA, an extremely asymmetric car-following behavior can be observed. Here the threshold of changing speed is equal to the minimum time headway Δt. Thus, the time headway between two vehicles can never be smaller than the threshold of changing speed. In this car-following system, a vehicle can never over- and/or under-steer his speed. The oscillating motion which is the basic property of a car-following system in real-world cannot be realized (cf. Figure 10). Therefore, the vehicles in a convoy form only an incompressible system which is always very rigid and stable.

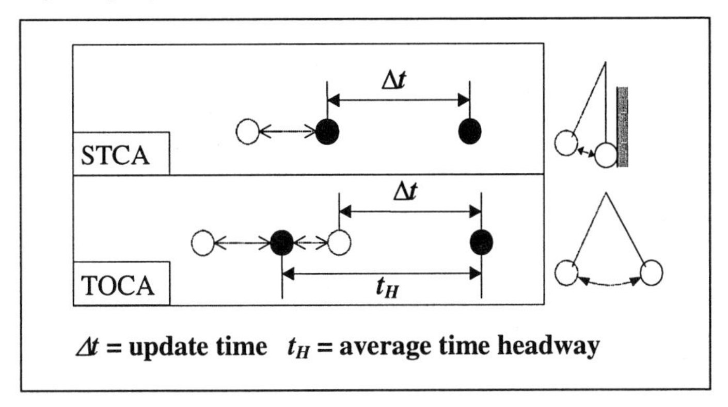

Figure 10: Car-following behaviors in STCA and TOCA.

To solve this problem, a new model (time-oriented CA, TOCA) is introduced that uses a more elastic but still simple approach to the representation of car-following. The TOCA has similar rules as the STCA. Only the distance behavior (car-following) between vehicles is modified. A new parameter t_H (corresponding to the average time headway in a convoy) is introduced that represents the threshold for changing speed (causing acceleration or deceleration). The pseudo-code of the rules can be rewritten as

(1) if (gap > $v \cdot t_H$) and ($v < v_{max}$) then $v = v + 1$ with p_{ac} (The speed is increased by 1 with the probability p_{ac} if the time headway to the vehicle in front is larger than t_H. An average acceleration ratio with the value p_{ac} is resulted.)

(2) if (v > gap) then v = gap

(3) if (gap < $v \cdot t_H$) and ($v > 0$) then $v = v - 1$ with p_{dc} (The speed is reduced by 1 with the probability p_{dc} if the time headway to the vehicle in front is smaller than t_H. An average deceleration ratio with the value p_{dc} is resulted.)

(4) $x = x + v$

Here the rules (1) and (3) are different from the rules of STCA. These two rules introduce a time threshold t_H that initializes changing of speed combined with random elements. If one chooses $t_H > \Delta t$ the car-following system becomes elastic and a realistic oscillation can be utilized (cf. Figure 10). The parameters p_{ac} and p_{dc} represent the average acceleration and deceleration ratio of the vehicle population. The parameter p_{brake} in the STCA can be omitted.

As a result, two major shortcomings of the STCA in the microscopic level can be partly improved: the acceleration and deceleration ratio have independent values (p_{ac} and p_{dc}); the threshold for changing speed (= time interval initializing adjusting speed) can obtain arbitrary values ($t_H > \Delta t$). Consequently, the maximum flow ratio in the opening phase C_{open} can be set smaller than the maximum flow ratio in the closing phase C_{close}. This leads to the following macroscopic properties of the TOCA:

1. Capacity drops within the range of capacity become possible.
2. Non-stable convoy dynamics and therefore breakdowns in a open pipeline system becomes possible.
3. More realistic speed-flow relationship.

These macroscopic characteristics agree with the real-world traffic flow on motorways very well.

In the same manner, also the rules for changing lanes can be modified into time-oriented rules (cf. Figure 11):

(1) if ($gap < t_{H,r} \cdot v_{max}$) and ($gap_{left} >= gap$) then desire-from-right-to-left.

(2) if ($gap > t_{H,l} \cdot v_{max}$) and ($gap_{right} > t_{H,r} \cdot v_{max}$) then desire-from-left-to-right.

(3) if ($gap_{back} >= t_{H,b} \cdot v_{back}$) then possibility-of-changing-lane.

(4) if ($v_{right} > gap_{left}$) then $v_{right} = gap_{left}$ (prohibition of passing on the right).

The parameter $t_{H,l}$, $t_{H,r}$ and $t_{H,b}$ are the time thresholds that determine the lane-changing behavior. They can be set to the average time headways in a convoy. They can also take different values representing the behavior of change-to-left, change-to-right and look-behind. The parameters p_{l2r} and p_{r2l} in the STCA can remain.

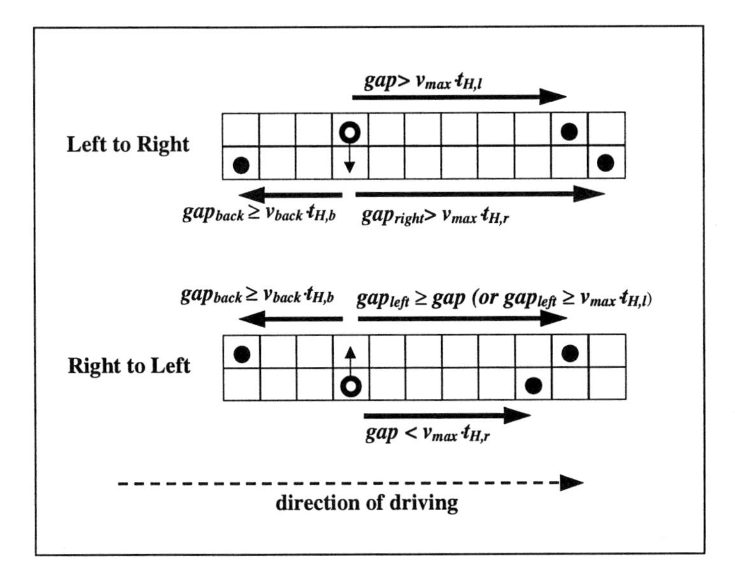

Figure 11: Principle of lane-changing in TOCA.

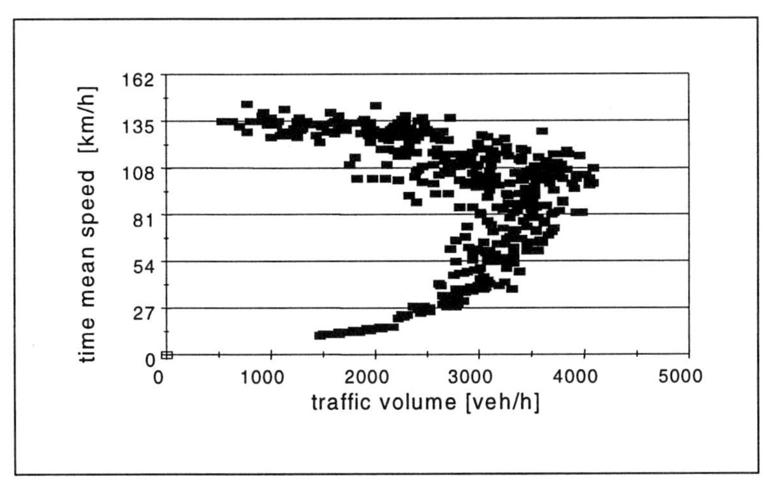

Figure 12: Speed-flow relationship from the new TOCA.

The new rules of the CA result into a much better agreement with the measurements. Figure 12 shows the results of the TOCA with the new rules for the same input flow pattern like the field data. Compared to the STCA an obvious enhancement can be observed (cf. Figure 7 and Figure 8). To obtain the speed-flow diagram in Figure 12 the parameters in Table 4 were used.

As a summary, the properties and their assessments of STCA and TOCA are listed together in Table 5. The TOCA modified the shortcomings and retains most of the advantages of STCA.

car-following behavior			lane changing from left to right			lane changing from right to left		
p_{ac}	p_{dc}	t_H	$t_{H,r}$	$t_{H,l}$	$t_{H,b}$	$t_{H,r}$	$t_{H,l}$	$t_{H,b}$
0.9-1.0	0.9-1.0	1.1-1.2	4.0	3.0	2.0	2.0	1.4	1.4
-	-	sec	sec	sec	sec	sec	sec	sec

Table 4: Parameters of TOCA for the example.

	STCA	Assess-ment	TOCA	Assess-ment
car-following				
time	discrete	±	discrete	±
space	discrete	±	discrete	±
desired speed	discrete	–	discrete	–
acceleration and deceleration	∞	–	< ∞ *	+
min. possible spacing	= Δt	–	= Δt	–
number of parameter	1	+ +	1 to 3	+ +
threshold for changing speed	= Δt	– – –	$= t_H > \Delta t$	+ + +
flexibility	low	-	high	+
calibration	very easy	+ + +	easy	+ +
lane-changing				
number of parameter	3	+ +	1 to 6	+ +
threshold for changing lane	= Δt	– – –	$= t_{H,x} > \Delta t$	+ + +
flexibility	low	-	high	+
Operation				
bitwise implementa-tion	easy	+ +	easy	+ +
parallel computing.	possible	+ +	possible	+ +
Speed	very high	+ + +	high	+ +

Table 5: Properties and assessments of STCA and TOCA.

4 Conclusion

Traffic on streets and highways can be simulated efficiently by the simplified microscopic CA-model. A CA can be described with the main properties: discrete division in time and space; isolate driving behavior and updating rules. The most important feature of a CA is the possibility of parallel computing.

Although the so-called STCA from Schreckenberg and Nagel produces good results for modeling queuing systems, it is not accurate enough for modeling the car-following behavior in the real-world and is therefore not suitable for modeling the traffic flow on motorways. The STCA was comprehensively calibrated and validated.

The new TOCA presented here retains the structure of a CA and introduces some new time-oriented rules which represent the real-world car-following behavior more realistic. The rules can still be updated in parallel for any vehicles, so that the possibility for parallel computing on super computers is maintained. The new TOCA reproduces the macroscopic laws observed in real-world traffic much better than the STCA and should be recommended for simulating traffic flows in motion, especially for traffic flows on motorways.

Before the TOCA can be used in practice, the system parameters of the TOCA should be calibrated and validated with more field data for different types of motorways and two-lane rural highways regarding the macroscopic behavior of traffic flows, especially the speed-flow relationships and capacities for different driving conditions.

References

1. Brilon, W. and Ponzlet, M. (1995): Application of traffic flow models. in: D. E. Wolf, M. Schreckenberg, A. Bachem (Edts), *Proceedings of Workshop in Traffic and Granular Flow*, World Scientific, Singapore 1995.

2. Brilon, W. and Wu, N. (1997): Calibration and validation of Cellular Automaton. *Technical Report.* Institute for Traffic Engineering, Ruhr-University Bochum. 1997.

3. Heidemann, D. and Wimber, P. (1982): Typisierung von Verkehrsstärkeganglinien durch clusteranalytische Verfahren (Standardisation of traffic flow patterns over time by cluster analysis). Series *Strassenverkehrszählungen, No. 26*, 1982.

4. Helbing, D. (1995): High-fidelity macroscopic traffic equations. *Physica A*, 219: pp. 391-407.

5. Lighthill, M.T. and Witham, G.B. (1955): On kinematic waves, A theory of traffic flow on long crowded roads. *Proceedings of the Royal society*, London, Series A, Vol. 229.

6. Nagel, K. (1995): High-speed microsimulations of traffic flow. *Ph.D.-thesis,* University of Cologne. 1995.

7. Nagel, K., Schreckenberg, M. (1992): A cellular automaton model for freeway traffic. *J. Phys. I France*, 2:2221-2229.

8. Nagel, K.; Wagner, P. and Wolf, D. E.(1996): Lane-changing rules in two-lane traffic simulation using cellular automata: II. A systematic approach. *J. Phys. A*, in preparation.

9. Ponzlet M. (1996): Dynamik der Leistungsfähigkeiten von Autobahnen (Dynamic of motorway-capacity). *Ph.D.-thesis, Ruhr-University Bochum.*

10. Sparmann (1978): Spurwechselvorgänge auf zweispurigen BAB-Richtungsfahrbahnen (Lane-changing operations on two-lane carriageways). Series *Forschung Straßenbau und Straßenverkehrstechnik*, No. 263, 1978.

11. Wagner, P. (1995): Traffic simulations using cellular automata: Comparison with reality. in: D. E. Wolf, M. Schreckenberg, A. Bachem (Edts), *Proceedings of Workshop in Traffic and Granular Flow*, World Scientific, Singapore 1995.

12. Wagner, P., Nagel, K. and Wolf, D. E. (1996): Realistic multi-lane traffic rules for cellular automata. *J. Phys. A*, submitted.

13. Wiedemann, R. and Hubschneider, H. (1987): Simulationsmodelle (Simulation models) in Lapierre, R.; Steierwald, G. (Edts.): *Verkehrsleittechnik für den Straßenverkehr.* Springer-Verlag 1987.

14. Wolfram, S. (1986) (Edt): Theory and Applications of Cellular Automata. World Scientific, Singapore 1986.

Effects of New Vehicle and Traffic Technologies –

Analysis of Traffic Flow, Fuel Consumption and Emissions with PELOPS

H. Wallentowitz[1], D. Neunzig[1], and J. Ludmann[2]

[1]Institut für Kraftfahrwesen Aachen, Steinbachstr. 10, 52074 Aachen, Germany

[2]Forschungsgesellschaft Kraftfahrwesen Aachen, Steinbachstr. 10, 52074 Aachen, Germany

This contribution presents the conception and application possibilities of the program system PELOPS that is being applied in the Northrhine-Westphalia Research Cooperative „Traffic Simulation and Impacts on the Environment", in order to predict the influence of new vehicle- and traffic concepts on the traffic flow and emissions. Apart from the presentation of the vehicle-orientated traffic simulation program PELOPS single focal points of projects of the Institut für Kraftfahrwesen Aachen (ika) in the frame of the FVU are shown and concrete application examples for PELOPS are demonstrated. Particular attention is paid to the calculation of emissions of traffic and to an explanation of necessary actions.

1 Introduction

Today's traffic finds itself in the middle of different, partly each other excluding requirements. The economically important transport of people and goods has to be as efficient and safe as possible but at the same time restrictive conditions concerning the reduction of pollution emissions and traffic noise exist. The consideration of both requirements places always higher demands on planer and developers of new traffic technologies. In recent years, simulation has been established as powerful tool in this respect. Due to this reason, a closed chain of simulation models were developed and applied in the Nordrhine-Westphalia Research Cooperative „Traffic Simulation and Impacts on the Environment". These simulation models are able to cover all questions from traffic flow up to direct effects on the environment. For a prognosis of the consequences of new vehicle and traffic technologies for traffic throughput, traffic safety and pollution emissions a particularly high-soluting simulation must be employed. The simulation tool PELOPS was especially developed for this purpose at the Institut für Kraftfahrwesen Aachen (ika).

In this lecture, first, it will be shown by means of PELOPS' conception, what kind of requirements are to be demanded of the vehicle-modelling and thereby time- and space-regarding highly soluting traffic simulation and which possibilities present themselves by their application for the planning and development of new traffic technologies. One prerogative for the application of traffic simulation is that the employed models do really reproduce the traffic flow absolutely accurately. In order to prove that, the NRW-FVU started a field study in the summer of 1996, by which all relevant data for the verification of the employed simulation models were to be measured. Ika analysed the vehicle license plat's numbers registered during the field study, in order to get exact information concerning the composition of the vehicle population. It was the aim to register the actual vehicle composition in respect to observed pollution limit values and the technical standard of emission control. This is absolutely necessary, in order to make exact statements about the actual emission of the traffic. Results of this evaluation will be exemplary shown in this lecture. Finally, a presentation of the reality truth of PELOPS-results, the presentation of concrete PELOPS applications in the field of traffic technology as well as calculations of fuel consumption and emissions of road traffic will be given.

2 The Program System PELOPS

PELOPS was developed at the Institut fuer Kraftfahrwesen Aachen in cooperation with BMW AG. The idea of PELOPS lies in a combination of detailed submicroscopic vehicle models with microscopic traffic-technical models, which enable an investigation of the longitudinal vehicle behaviour as well as an analysis of traffic flow. The advantage of this method lies in the fact that all interactions between driver, vehicle and traffic can be taken into consideration. The realisation of such a complex tool would not have been realised without the rapidly increasing power of computers, so that today a calculation of traffic situations can be executed by PELOPS in real-time.

In its conception PELOPS is orientated at the main elements of the traffic system - stretch/environment, driver and vehicle [1]. In a modular program structure the named elements are modelled and limited by defined interfaces (see fig. 1). The stretch model allows a detailed description of the influences of a stationary traffic environment. The course of the road in horizontal and vertical direction by radius and transitions is indicated as well as the number and the width of the lanes. In addition, geometric data, traffic signs as well as environmental conditions can be given by defined parameters such as wetness, slippery surface etc.

The actual traffic conditions for a vehicle result from the number of surrounding vehicles and their distance and velocity. In order to depict certain behaviours in traffic or to reproduce given driving cycles, selectable speed profiles can be given to the driver-vehicle-units.

The vehicle model bases on the „Cause and Effect-Principle" [2]. A calculation of the propulsive power starting from the engine operation point over clutch, drive and differential up to the tires is done, where the propulsive power is then balanced with the road resistance. The operation point is changed by an alteration of the engine torque (cause).

From the thereby caused acceleration and speed change result engine RPM (effect) under consideration of the drive-line elements. As transmission the conventional hand-shifting as well as automatic drive are implemented. For commercial vehicles a retarder can be additionally depicted in the drive-line. Only such a detailed depiction of the vehicle under application of the Cause-Effect-Principle enables an investigation of control elements such as ABS and ACC [3].

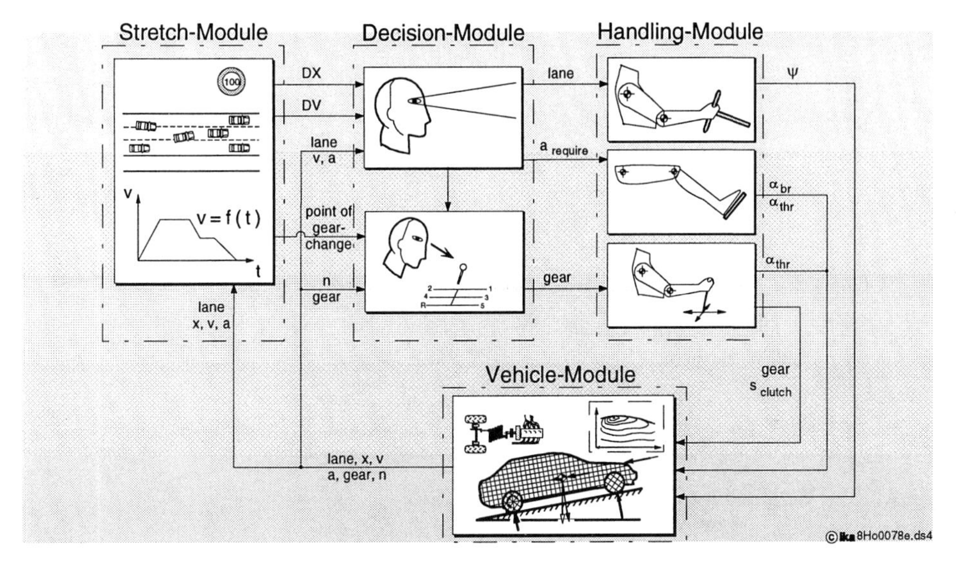

Figure 1: Basic Structure of PELOPS.

The link between vehicle- and traffic simulation is represented by the driver model. It is divided into a decision and a handling model. In the decision model the parameter of local driving strategy is determined by the actual driving conditions and the vehicle environment. The parameters of local driving strategy are the desired acceleration, the driving lane and also the gear lever. In the handling model these parameters are converted into vehicle controls such as position of acceleration pedal, brake etc.

For a determination of the driver's behaviour PELOPS works with a psycho-physical following model that grounds on the work of Wiedemann [4]. This model was adjusted and significantly further developed for it's application in PELOPS [5, 6]. Because the follow-model does only consider the reaction to vehicles driving in the same lane, a tactical behaviour was additionally developed that creates a realistic driver behaviour in respect to the stretch course and in case of multilane roads. This tactical behaviour includes, for example, the reaction to slow vehicles on the neighbouring lanes, to different traffic signs and intersections as well as to a situation-depending lanechanging model.

For example, the parameterisation of the lanechanging model is adjusted to different lanechanging situations, in order to be able to depict the different behaviours of drivers. Depending on the driver's satisfaction with his own lane a so-called lanechanging desire is determined. In case the space on the neighbouring lane is big enough, the driver model initiates a lane change (see fig. 2). The size of the needed lane changing space depends on the actual driving situation, the difference in speed between the vehicles and the individual safety need.

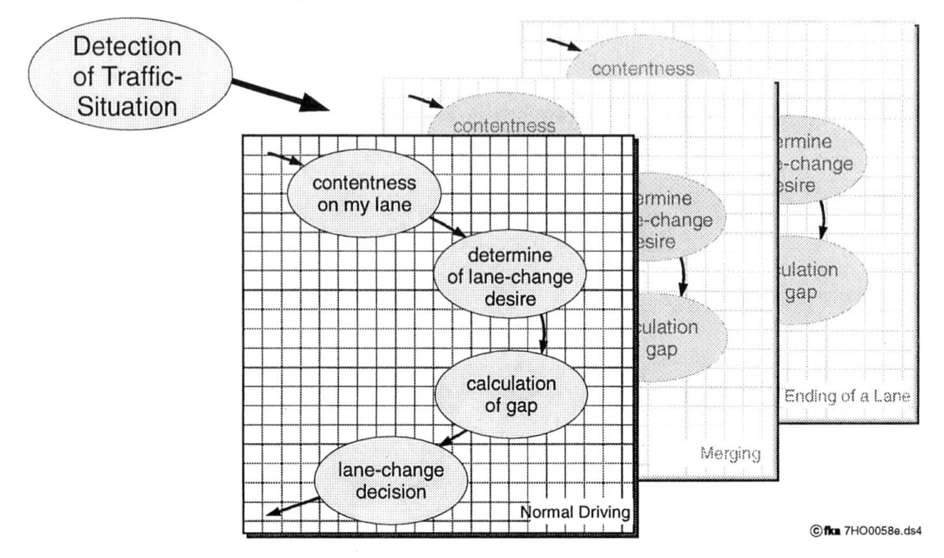

Figure 2: Design of the Lane Changing Model [7].

PELOPS does thus employ very high-soluting models, in order to depict the traffic situations in the highest quality. However, the model is only applicable, if it is truly capable to deliver realistic results and if it disposes of sufficiently accurate input data. Due to this reason, the necessary data for an examination of the simulation quality was registered and based on this, extensive validation work was done in the frame of the NRW-FVU. Furthermore, extensive input data for all different kinds of traffic scenarios were determined. The main data source was in both cases the field study of the NRW-FVU.

3 Field Study and Validation

In June 1996, the field study of the NRW-FVU took place on the Nordrhine-Westphalia highway net in a triangle between Aachen, Cologne and Duisburg. In the course of the project this field study was extended at ika by single measuring days. Goal of the study was to get input data for the single simulation models as well as to create a data basis for a comprehensive validation. In a first evaluation block the input data needed for PELOPS was evaluated by ika. This evaluation can be divided into two parts. At first, measuring quantities were identified that describe the driver behaviour depending on traffic situation and stretch. Fig. 3 exemplary shows the evaluation of the driven velocities on the highway A4 Cologne-Aachen at the height of the access point Dueren.

The measuring took place during the morning rush hour. In the investigated stretch the speed limit of 100 km/h is indicated by means of changeable traffic signs. In fig. 3 the lane of the access point Dueren is named lane 1, the right and left lane are named lane 2 and lane 3. It becomes apparent that especially on the left lane number 3 less than 50% of the vehicles keep the speed limit. This low acceptance of the speed limit results from the special characteristics of the changeable traffic signs on the A4. During the rush hour the speed limits, indicated by succeeding changeable traffic signs, change without this being always understandable for the drivers. The resulting speed behaviour, described in Fig. 3, must be realistically depictable by such a high-soluting traffic simulation tool as PELOPS, in order to reproduce the real traffic throughput and especially the emission behaviour accurately.

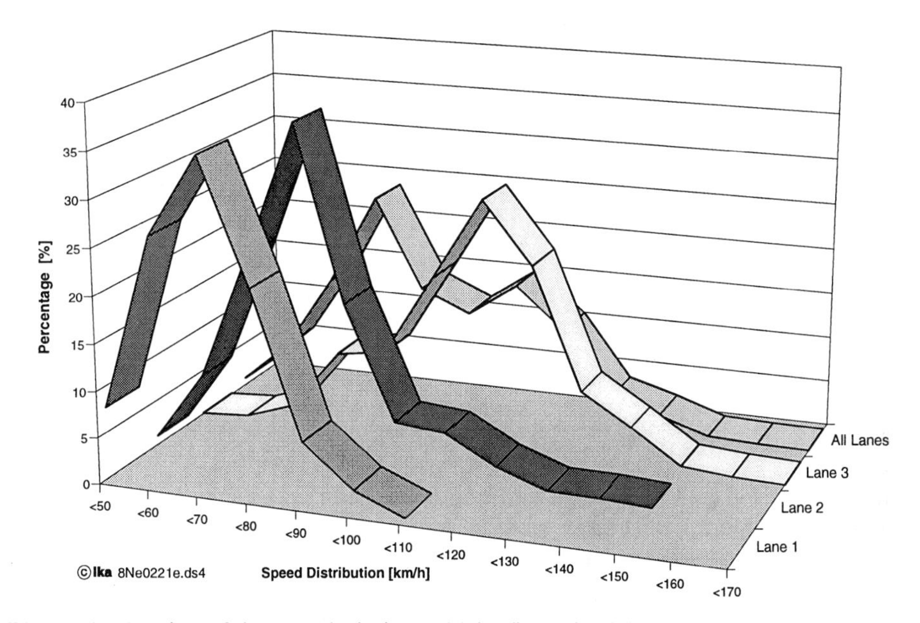

Figure 3: Keeping of the Speed Limit to 100 km/h on the A4.

A further important quantity for the determination of emissions on a stretch section is the composition of the vehicle population that is driving on this section. The relevant quantities are on the one hand the valid pollution limit class, the single vehicles belong to (f. e. EURO 2) and on the other hand the quantities vehicle type, year of construction, capacity, engine type and -power. In order to determine these quantities the license plates of the vehicles were registered and handed over to the Kraftfahrt-Bundesamt for evaluation. The data could then be compared to available statistics concerning vehicle stock. Apart from the official registration statistic the data bank „Handbuch fuer Emissionsfaktoren des Straßenverkehrs“ [8] (Handbook for the Emission Factors of Road Traffic) was also consulted (in the further course of the lecture named „handbook“). The handbook distinguishes specific vehicle distributions for different road types (f.e. city - overland road and highway) that were also implemented into the emission calculations.

Fig. 4 compares the vehicle distribution for different emission classes between the evaluation of ika's field study and the vehicle distribution for the highway used by the handbook. The field study data of the NRW-FVU was collected on the highway ring Cologne-West (A1) in northern driving direction. The two bright lines in front are valid for Diesel-engined vehicles, whereas the two dark lines in the back describe the distributions of gasoline-engined vehicles.

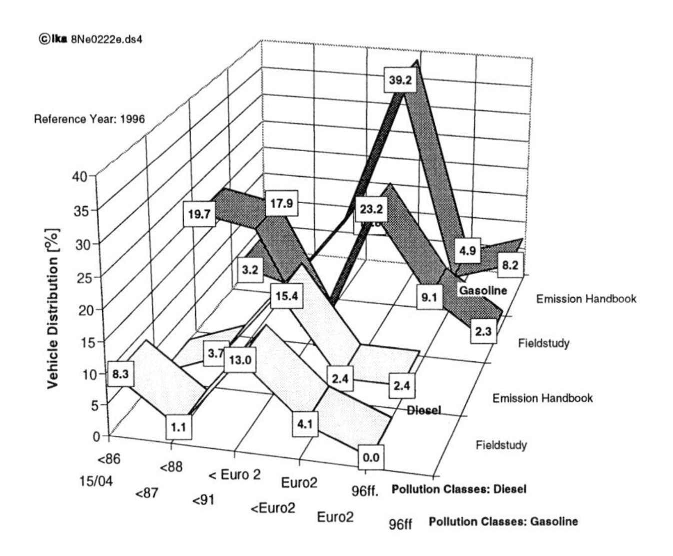

Figure 4: Vehicle Distribution on Highway: the Field Study and the „Handbook".

It becomes apparent that a high correspondence for Diesel-engined cars exists between the handbook and the field study with the exemption of the pollution class „before 1986 (<86). However, this deviation is not neglectable, because particularly these vehicle show an especially high pollution emission. In case of the gasoline-engined cars a significantly greater deviation between the handbook and the field study can be established. In the field study considerably more vehicle of the pollution classes „ECE 15/04" and „<87" were counted. In contrast to that the share of „Before EURO 2"-vehicles is about 16%-points lower. Overall it can be established that in the handbook a substantially less polluting vehicle population is assumed than it was actually found at the investigation point Highway A1 (Ring Cologne-West). The distribution after the handbook would therefore lead to significantly lower pollution emission than in reality.

Due to their higher accuracy the results of the field study serve as basis for the simulation of the highway ring Cologne-West by means of PELOPS. Apart from the selection of realistic input data for the simulation it has to be ensured furthermore that the used simulation models deliver realistic results. In case of PELOPS this is done by means of the highway ring Cologne-West. Fig. 5 shows in a fundamental diagram minute values average speed and traffic density measured over a period of 24 hours. In comparison to that three PELOPS simulation for free traffic, collapsing traffic and traffic jam are presented. The simulations cover thereby always a period of up to one hour.

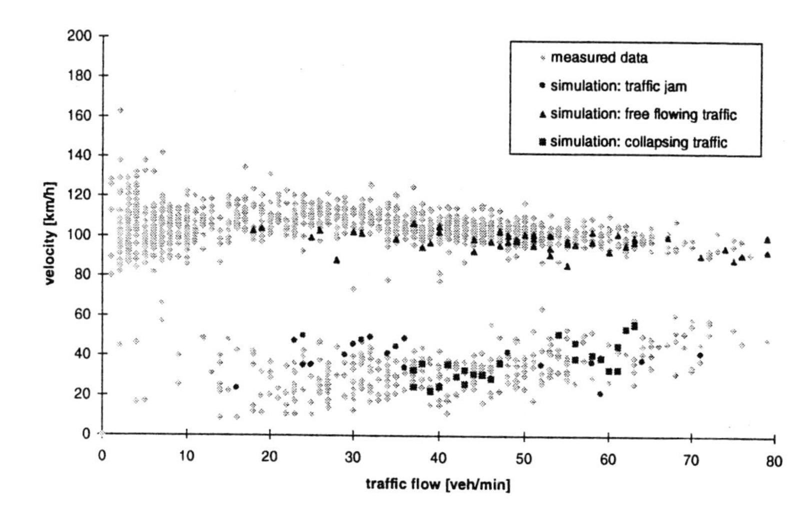

Figure 5: Macroscopic Comparison between Measuring on the A1 and Simulation.

Overall, a very high correspondence between measuring and simulation is shown so that the macroscopic effects of traffic can be realistically reproduced by PELOPS.

Next to the reproduction of so-called macroscopic comparison quantities such as traffic density and average speed it has to be proven that realistic results are also obtained on a microscopic level. Fig. 6 (see above) shows by means of the microscopic quantities distance and relative speed in respect to the preceding vehicle the high microscopic reproduction quality of PELOPS. A succeeding drive in city traffic following a randomly selected vehicle was simulated. Thereby, no special parameterisation was selected by PELOPS, but it was calculated with standard setting for an „average“ driver. Overall, the powerfulness of PELOPS could be proven within the frame of the NRW-FVU. A particular advantage of PELOPS lies in the very high reproduction quality with simultaneous highest time- and space-regarding solution. In the following chapter it will be shown by examples what kind of application possibilities PELOPS offers due to the described advantages.

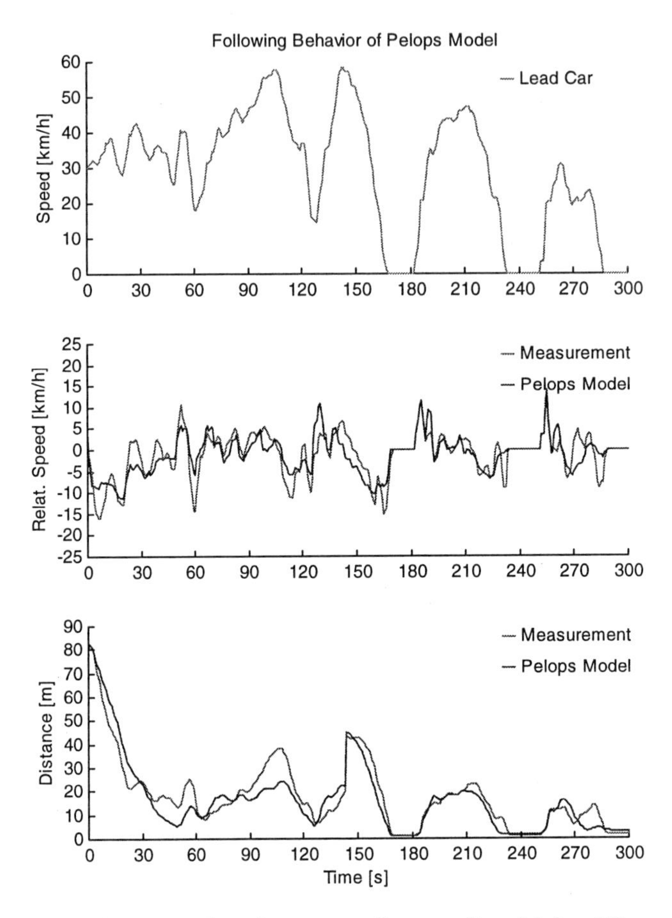

Figure 6: Microscopic Comparison by means of succeeding driving [9].

4 Applications

Fig. 7 shows a general overview of PELOPS application groups and typical applications. The spectrum of the applications begins with the development of vehicle control systems such as the so-called „Intelligent Cruise Control" (ACC, Adaptive Cruise Control) up to the investigation of traffic measures concerning their effects on traffic flow and emissions.

Traffic courses on highways up to city traffic can be comprehensively analysed and optimised by PELOPS. Thereby, the examination of different knotpoint forms (f.e. intersections, roundabouts) is possible with conventional or intelligent traffic light control. These possibilities were employed, in order to design a traffic-depended traffic control for a highly frequented access road in the Aachen suburban area. A one-laned section with a succession of three traffic-light controlled intersections was investigated. In the original version the traffic lights were controlled by four signal programs, depending on the day-time. Fig. 8 presents the results of the executed optimisation works with PELOPS. Apart from improved traffic throughput and travelling time, also the toxic emissions could simultaneously be reduced. This is demonstrated in Fig. 8 by means of the emission- and consumption reduction.

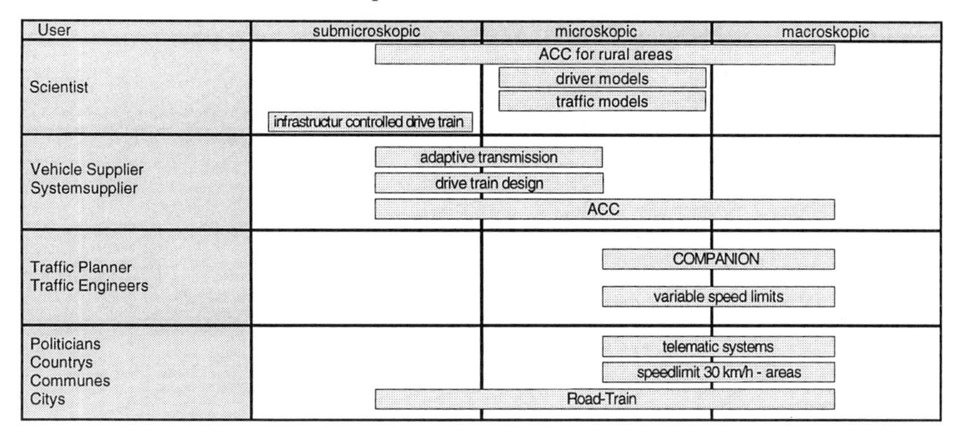

Figure 7: Examples for the Application Spectrum of PELOPS.

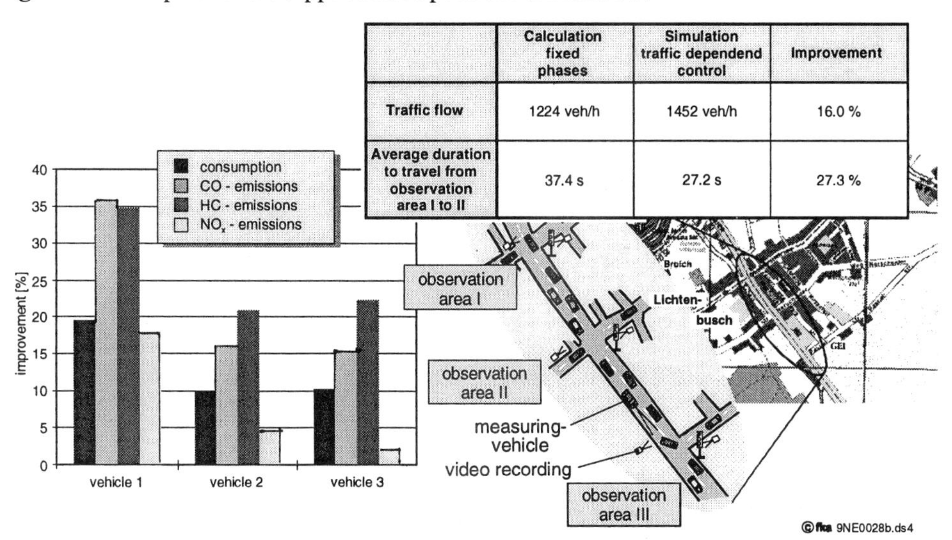

	Calculation fixed phases	Simulation traffic dependend control	Improvement
Traffic flow	1224 veh/h	1452 veh/h	16.0 %
Average duration to travel from observation area I to II	37.4 s	27.2 s	27.3 %

Figure 8: Results of the Design of an Intelligent Traffic Light Control.

Illustrated by the example of the roadside driver warning system COMPANION it can be shown, how measures for the improvement of traffic flow can be worked out with PELOPS. COMPANION is currently tested on the A92 north of Munich in co-operation between the State of Bavaria and BMW. In case of an accident or a potential dangerous traffic obstruction COMPANION warns the driver by blinking roadside beacons. The reaction of the driver to the blinking beacons was implemented into PELOPS, in order to provide detailed simulations for the test section of the A92. By the use of COMPANION the speed level is reduced before the accident site and the traffic flow becomes thereby significantly smoother and less dynamic [10].

Through the reduction of the different speeds between normal and cautious driving the probability of accidents is clearly reduced. At the same time the possible speed collapse can be softened by COMPANION and thereby the throughput increased. Fig. 9 shows the influence of COMPANION in case of a lane reduction from 3 to 2 lanes.

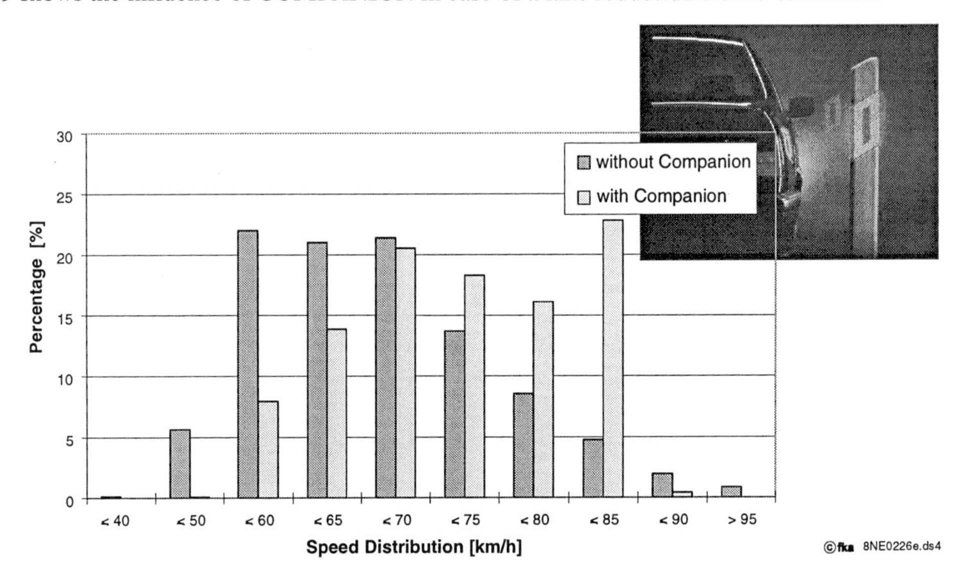

Figure 9: Speed 100 m behind a lane reduction from 3 to 2 lanes - with and without COMPANION.

5 Conclusion

Today's automotive development does not consider the vehicle as an isolated item, but instead as part of the whole system man, environment and traffic. For an analysis of the vehicle's influence on this whole system as well as for an analysis of the interaction between vehicle, man and environment the submicroscopic traffic simulation tool PELOPS is developed and applied at the Institut fuer Kraftfahrwesen Aachen. In this paper PELOPS was presented, and single results of the validation works in the frame of the NRW-FVU were explained for models used in PELOPS. The completion of this lecture was formed by selected application examples of PELOPS that made clear what kind of application possibilities offer themselves for the vehicle-orientated traffic simulation.

References

1. R. Diekamp, Entwicklung eines fahrzeugorientierten Verkehrssimulationsprogramms, Dissertation am Institut für Kraftfahrwesen, Aachen, 1995.

2. W. David, Modulares Simulationsprogramm zur Auslegung und längsdynamischen Untersuchung von Kfz-Antriebssystemen, Dissertation am Institut für Kraftfahrwesen, Aachen, 1992.

3. J. Ludmann, M. Weilkes, Entwicklung, Analyse und Bewertung von PROMETHEUS Konzepten unter be-sonderer Berücksichtigung von Autonomous Intelligent Cruise Control mittels Simulation, Abschlußbericht zum Eureka Verbundprojekt, Aachen 1995.

4. R. Wiedemann, Simulation des Straßenverkehrsflusses, Schriftenreihe des Institutes für Verkehrswesen der Universität Karlsruhe, Heft 8, Karlsruhe, 1974.

5. J. Ludmann, D. Neunzig, N. Weilkes, Traffic Simulation with Consideration of Driver Models, Theory and Examples, Vehicle System Dynamics, Vol. 27, 5-6, pp. 491-516, 1997.

6. J. Ludmann, Beeinflussung des Verkehrsablaufs auf Straßen - Analyse mit dem fahrzeugorientierten Verkehrssimulationsprogramm PELOPS, Dissertation am Institut für Kraftfahrwesen, Aachen, 1998.

7. J. Ludmann, A. Hochstädter, D. Neunzig, M. Weilkes, Assessment of Advanced Vehicle Control Systems with th Vehicle Oriented Traffic Simulation Tool PELOPS, SAE - 1998 Future Transportation Technology Conference, Costa Mesa, California.

8. Handbuch Emissionsfaktoren des Straßenverkehrs, Umweltbundesamt Berlin, 1995.

9. D. Manstetten, W. Krautter, T. Schwab, Traffic Simulation Supporting Urban Control System Development, Proc. of the 4th World Congress on Intelligent Transport Systems, 21.-24. October 1997 in Berlin, Germany.

10. D. Ehmanns, Modellierung des Fahrerverhaltens bei verschiedenen Verkehrszuständen zur Analyse von Fahrerassistenzsystemen, Diplomarbeit, Institut für Kraftfahrwesen der RWTH Aachen, 1996.

Traffic Simulation for the Development of Traffic Management Systems

W. Krautter, D. Manstetten, and T. Schwab

Robert Bosch GmbH, Corporate Research and Development, Dept. FV/FLI - Information and Systems Technology, P.O. Box 10 60 50, 70049 Stuttgart, Germany

This paper describes the industrial application of traffic simulation at Robert Bosch GmbH. The role of traffic simulation in the design process of traffic management systems and the way of working of Bosch R&D are presented. The main features of Bosch´s simulation environment ARTIST (Advanced Research Tool for Indoor Simulation of Traffic) are described. Sample applications in the area of freeway management systems and urban traffic light control demonstrate how traffic simulation is currently used in the product development process. ARTIST´s flexibility allows the usage of domain specific traffic models that are best suited to the specific task.

1 Introduction

Traffic simulation may serve as a powerful tool in the design process of traffic management systems. It is possible to get a first estimation of the function and the impact of planned measures long before the real installation of a system is done. All design phases from the "overall product design" over the "layout of the system" to the "optimization" can be supported by a simulated model of the product and the traffic environment.

Like all simulation approaches it is possible to compare and assess different variants of a product already in a very early state of the design process. This makes the specification task much more efficient and transparent since many impacts may be easily demonstrated to customers and decision makers in advance.

Since the cost of traffic simulation is not very high, it may also serve as a cost-effective enhancement to field trials. It is possible to initially resolve all questions that may be answered by a computer based modeled scenario, and leave the expensive field trials to the remaining tasks that can only be done with a real world installation. In some cases when overall system behavior is the main focus of the experiments, traffic simulation may even serve as a substitute for field trials. Specific traffic scenarios are often much easier to reproduce in a simulated computer model than in a real world environment. Another benefit is the possibility to study scenarios in the laboratory without any safety risks. This is essential for studying real break-down scenarios for traffic management systems (like accidents and jams).

In the near future traffic simulation will also be used more and more for product marketing. To demonstrate product advantages an animated "live" simulation is very intuitive and requires less equipment than real hardware setups of the product.

2 Traffic Simulation at Bosch

In 1993 Bosch set up a traffic simulation group inside the corporate research and development to study the interaction of driver-vehicle-traffic in the area of typical Bosch products like „driver assistant systems" and „traffic management systems" (see Tab. 1). This group acts as a central service provider for the product divisions inside the company. The mission within the company is:

- Providing and maintaining a simulation environment that is powerful and flexible enough to study different products and scenarios.
- Cooperating with product divisions on specific projects. The focus of the work depends on the problem and may be
  - modeling of new product components and new traffic models,
  - analysis & evaluation of simulation results,
  - or animation & visualization of product benefits for external presentation.
- Serving as an interface to the research community in the field of traffic and driver models and their applicability to product related questions.
- Participating in publicly funded projects (e.g. MOTIV, see [1]) together with other companies to develop a shared framework for using traffic simulation and interpreting the results.

Driver Assistance	Traffic Management
Adaptive Cruise Control	Freeway Management Systems
Collision Avoidance	Traffic Lights in Urban Areas
New Brake Systems	Parking Management
Navigation Systems	Priority for Public Transportation

Table 1: Typical Bosch product fields for traffic simulation.

Until now traffic simulation was used for products in the domains adaptive cruise control (ACC), new brake systems, freeway management system (FMS), and traffic lights in urban areas.

3 ARTIST - Bosch´s Simulation Environment

To serve the broad variety of applications, Bosch developed its own traffic environment ARTIST (Advanced Research Tool for Indoor Simulation of Traffic) to be used as an in-house tool [2]. In contrast to most other traffic simulation programs, ARTIST does not come with a predefined traffic or driver model. It is an event-oriented simulation environment with features adapted to the specific needs of traffic simulation. A powerful modeling language allows the definition of a rich set of re-usable model libraries. This approach is very similar to control engineering tools like SIMULINK [3] or MATRIXx [4]. The general architecture of ARTIST is shown in Fig. 1.

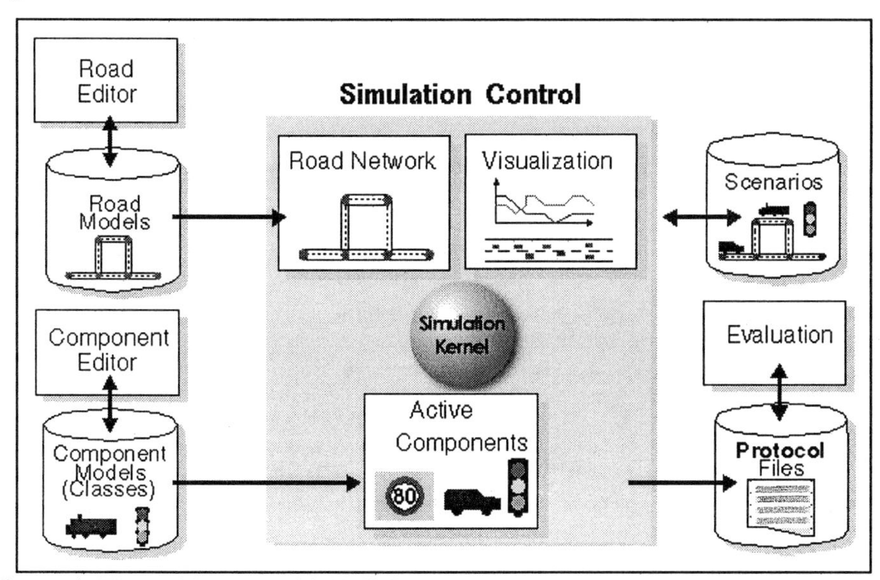

Figure 1: The architecture of ARTIST.

A road editor allows the definition of complex road models that are stored in a road database. The component editor serves to build up component classes for the model library that may be used by the simulator. Running simulations may be stopped and stored in the scenario database so that the simulation may be restarted without any losses from a defined starting point. Furthermore, ARTIST can produce configurable protocol data to external files.

The main design features of the traffic simulation environment ARTIST are:

- Reusability of configurable road networks with attributes of nodes, streets, and lanes (e.g. geometric information, slope, friction, traffic streams crossing intersections, right of way rules, etc.).
- Simulation kernel with a general approach suitable to multi-scope applications. It is also possible to simulate with macroscopic or microscopic traffic models.

- High-level object-oriented modeling language to define the behavioral description of active components in a simulation (e.g. driver-car units, car generators, traffic lights, sensors etc.). This language also provides a foreign function interface to connect object code directly into the simulation. This feature was heavily used in simulations for ACC where we linked in precompiled C code from the car's control unit.

- Reusability of models. The models are structured in a hierarchical way where each sub-component is stored in a library. Fig. 2 gives an example of a model for a driver-car unit of an ACC car with different microscopic following models that may be switched during simulation.

- Interaction and flexibility during runtime is one of the main features of ARTIST. It is possible to create new cars or traffic elements on the fly and to modify the parameters of each active traffic element (cars, controls, sensors, traffic signs, etc.) during the simulation.

- Animation and visualization is also available in different adaptable ways. A two-dimensional animation of the traffic scenario shows the location and selectable attributes of all vehicles and other objects on the road network. It is also possible to monitor the internal state (variables and parameters) of the traffic elements with configurable gauges, histograms etc., interactively during the simulation.

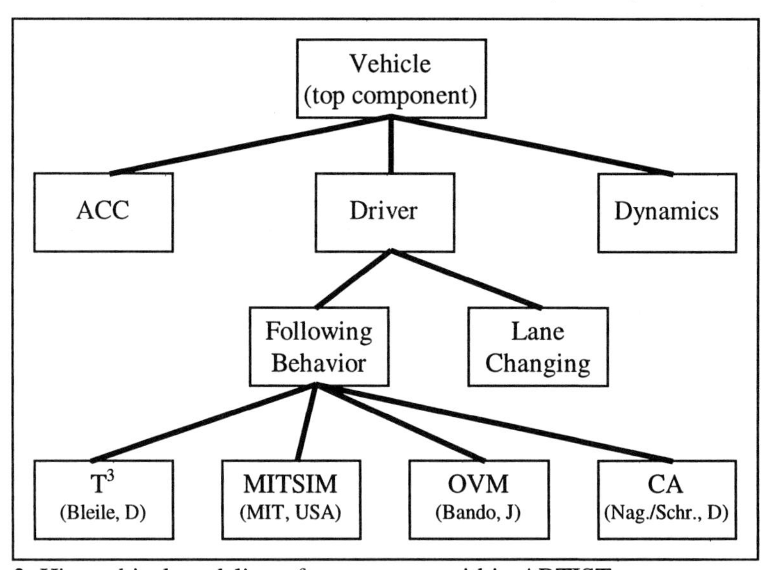

Figure 2: Hierarchical modeling of components within ARTIST.

4 Application of Traffic Simulation for Traffic Management Systems

In this chapter we will give an overview of some applications of traffic simulation that were done by Bosch in the product fields of freeway management systems and traffic lights in urban areas.

4.1 Application for Freeway Management Systems

Freeway management systems (FMS) are already widely installed to harmonize traffic flow and to reduce the number of accidents. The system functions as follows. Traffic is measured using induction loops or microwave detectors and the collected data is then sent to a traffic computer. The computer analyses the situation and issues appropriate speed limits and overtaking bans. The information is presented to drivers by means of variable traffic signs.

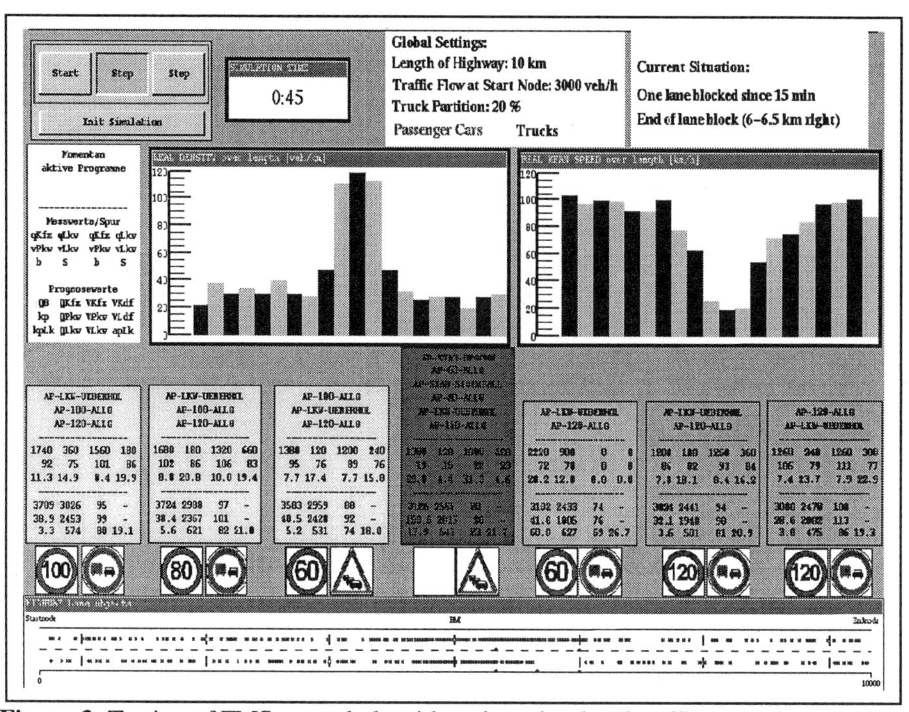

Figure 3: Testing of FMS control algorithms in a simulated traffic scenario.

The control programs currently used in FMSs are based on extensive empirical knowledge and usually require long, drawn out adjustment procedures before they can be incorporated into existing systems. The underlying control modules of the software libraries for Bosch's commercial FMS were tested by traffic simulation in different traffic scenarios. With this library Bosch is ready to reuse a set of well-proven algorithms when setting up a new FMS on a specific highway. Fig. 3 shows a screenshot of an ARTIST scenario where a traffic engineer is able to study and improve his underlying control algorithms in a complete virtual traffic setup. He is able to inspect the actual traffic situation in a live simulation, the measurements that were transmitted by the local sensors, and the interpretation and activation of programs that are changing the variable traffic signs.

Simulations can play a significant role in various phases of FMS planning and testing. At the design stage, they can be employed to determine the optimal measurement and display point spacings. The control systems can then be adapted in the initial stages of implementation using simulations, taking driver reactions to the signs into account. Consequently, the adjustment phase for the system is considerably shortened. Furthermore, critical traffic situations are easy to set up and reproduce using simulations, whereas they only occur sporadically in reality. However, it is just these situations that are crucial to the assessment of the system's quality.

During software development for the traffic control center for highway A94 (east of Munich, Germany), ARTIST was used for extensive software testing and even for performance verification in cooperation with the customer. The traffic control program has to signalize a 14 km highway segment, and in addition to conventional freeway management systems with variable signs on the highway itself, there is a ramp metering control added on the entry nodes from the fair area Riem. The system was put into operation in April 1998.

For early SW testing a complete road model of the equipped A94 with the seven intermediate entry and exit nodes per direction was modeled with ARTIST's road editor. The sensors and signs were positioned on this road network, and traffic was generated according to given empirical data for the known traffic scenarios at A94. An interface between ARTIST and the software program for the traffic control center was set up as shown in Fig. 4.

Both software programs communicate through the same protocol (according to the TLS standard [5]) that is used for transmitting data from the local sensors on the highway, and for signal orders to the variable traffic signs. In this setup, ARTIST plays the role of the complete traffic environment for the control center with simulated traffic, sensors, and traffic signs. ARTIST sends data from the simulated sensors to the control center and gets back signal orders to change the variable signs. The visualization of the traffic and the signs is also done by ARTIST. With this connection a complete feedback loop for studying the correctness of the signalization and the impact on the traffic for scenarios given by the customer was possible long before any real hardware elements were installed on the highway.

Bosch works not only on adapting conventional control programs, but also investigates fundamentally new control concepts. E.g., the potential of using fuzzy logic for early detection of jams (out of the measured data coming from the sensors) is being assessed with simulations [6]. It is essential to try out new approaches of this kind using traffic simulations as fundamental defects cannot be ruled out from the beginning. In real world trials they could cause chaos on the highway.

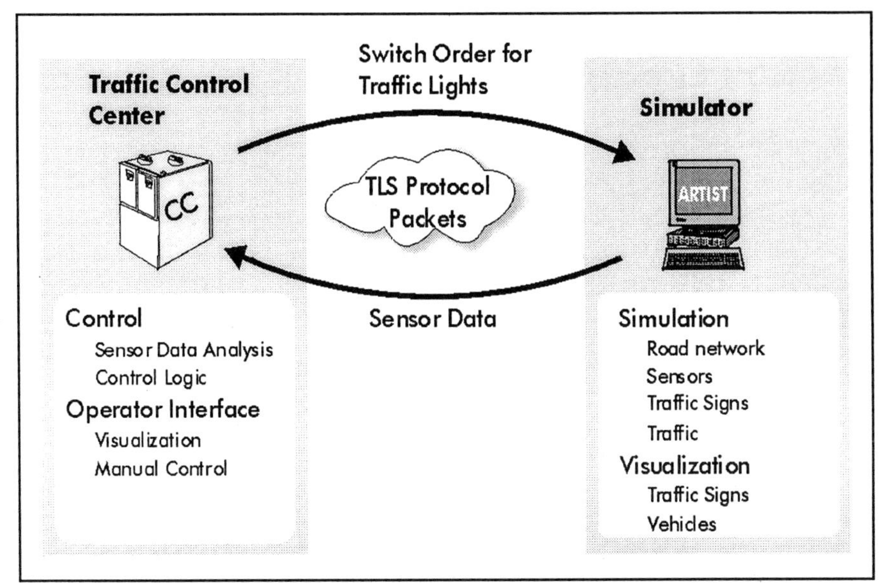

Figure 4: Interface between traffic simulator and traffic control center for A94.

4.2 Application for Traffic Lights in Urban Areas

Most existing traffic light control systems use a time-based control strategy for intersection signalization. Each intersection has a fixed cycle time split into green and red phases for the individual traffic lights. Depending on the time of day, these fixed time schedules may be changed but in general there is no adaptation to the real traffic situation. All adaptation is done in the design phase by defining the cycle times and green phases according to empirical data of the traffic that is expected for the intersection.

Newer approaches make use of sensors on incoming or outgoing lanes of each intersection to measure the traffic flow, speed, or occupancy on the lanes, and are able to adapt to real traffic automatically. This kind of adaptation may be an intersection-centered approach where each intersection is handled independently, or a more sophisticated global one where the signalization is coordinated to gain an optimum for the whole traffic network. In Bosch, traffic simulations are used to develop new adaptive control strategies based on these approaches.

Furthermore we also use ARTIST to evaluate the potential of the next generation of traffic sensors. These sensors can measure or generate additional data as compared to the data generated by current technology. For example, video sensors could provide data of traffic density on a lane. ARTIST is then used to estimate the value of such data in adaptive control strategies.

To study an existing real world problem, Bosch chose the CentrO area in Oberhausen, Germany as a good test site for the simulation. Bosch installed the park guidance system and the (up to now fixed-time based) traffic light control system for this large shopping and recreation area that covers app. 25 lane km and includes 21 signalized intersections. Fig. 5 gives an overview of the site.

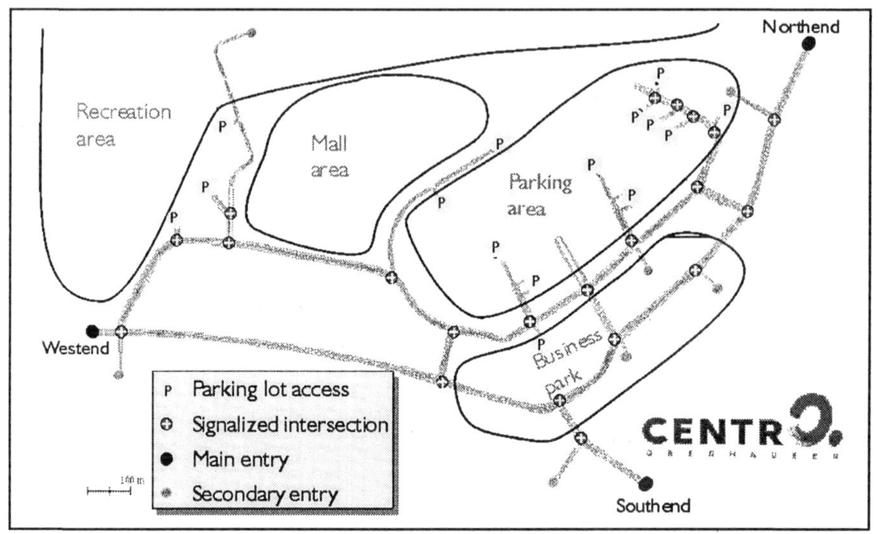

Figure 5: The CentrO area in Oberhausen, Germany.

In the CentrO area there are around 200 shops, different cinema centers, and 20 restaurants. It has a multi-purpose hall, a business center, and 600 apartments. Beside the shopping area is a recreation area with different attractions. There are 10 main parking garages with an overall capacity of 10,500 cars. The road network itself has three main entry/exit points (referenced as Westend, Southend, and Northend in Figure 5).

To study the effects of new control strategies for this network there is an ongoing research study at Bosch together with the Carnegie Mellon University (CMU) in Pittsburgh, USA. The goal is to design and implement an adaptive control program that is easily portable to different locations. The underlying idea is to use a multi-agent approach for the control. Each intersection control acts as an autonomous agent that has access to local sensors. Each agent knows only his direct neighbors and can exchange information with them. Global optimization shall be reached by negotiating local control strategies. The advantages over a centralized control scheme, where every data is collected at a central computer sending the switch orders to the traffic lights, are:

- Expandability: Adding new signalized intersections needs no software changes. Only the connection to neighbor intersections needs new configuration information.

- Simpler communication network: No networking from the central unit to each intersection is needed. Only direct neighbor connections are necessary.

To test and analyze this approach a connection similar to the one in Figure 4 between ARTIST and the control program was chosen. The whole CentrO area was modeled as an ARTIST road network with detailed descriptions of all the possible ways for vehicles to cross an intersection (the traffic streams of the intersection). A configuration file was used to equip the 21 signalized intersections with the necessary traffic lights and sensors. Each intersection could be switched to control the lights according to a predefined fixed-time control implemented inside ARTIST or to use an external adaptive control program.

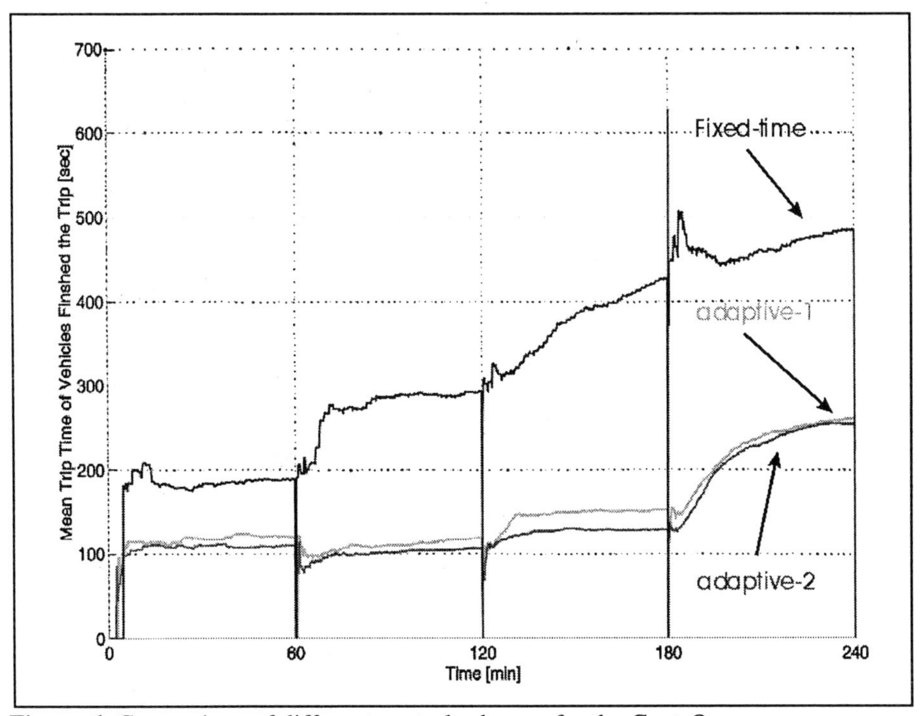

Figure 6: Comparison of different control schemes for the CentrO area.

Like in the application for the A94, ARTIST simulates the traffic and sends sensor data to the control program. The retransmitted signal orders are interpreted by ARTIST and the traffic lights are changed in the simulation accordingly. ARTIST is used to visualize the scenarios and also checks the correctness of the light settings (e.g. conflict checks for green traffic streams on intersections). ARTIST also generates general traffic measurement data (such as the delay time at an intersection, the mean trip time of the vehicles in the net) and saves them in an external file. Subsequently, other analytical tools are used with these data to evaluate the control algorithms.

One scenario that was studied is the typical traffic flow after a late cinema event. Most of the vehicles come out of the parking area in the northern part searching to leave the net at the three exit points Westend, Southend, or Northend. In reality this situation is one of the most critical that often leads to large jams even inside the parking garages. Figure 6 shows the comparison of mean trip time in this scenario for a fixed-time control and two different adaptive control schemes implemented up to now, demonstrating the clear benefit of the latter.

5 The Role of Traffic Models

To study the impact of traffic management systems on traffic flow, a closed-loop simulation is needed where the effects of the control are fed back to the traffic. To accomplish this the simulator needs a realistic traffic model covering all situations that are to be observed. To gain reasonable results by the simulation it is necessary to first choose the right traffic model for the task and to be well aware of the limitations of the particular model.

Bosch spent effort in studying and comparing many different traffic models. Most of the comparison was done by re-implementing the published algorithms in ARTIST and comparing their microscopic and macroscopic effects in different scenarios and with real traffic data. The results of the comparison are published in [7]. Tab. 2 gives an overview of the models implemented in ARTIST and the typical application where they were used.

Category		Basic Reference	Application			
			ACC	BS	FMS	TL
individual vehicles	Influenced by Direct neighbor	Wiedemann [9]	X	X		
		OVM (Bando) [10]			X	X
		Cellular Automata [11]			X	
		MITSIM [12]				
		T^3 (Bleile) [8]				X
	Influenced by multiple neighbors	Witte [13]	X	X		
	Influenced by macroscopic values	Bosch (unpublished)			X	
traffic flow	Influenced by macroscopic values	Cremer [14]			X	

Table 2: Traffic models implemented in ARTIST (ACC = adaptive cruise control, BS = brake systems, FMS = freeway management systems, TL = traffic lights).

As a result of the comparison it turned out that most of the microscopic car-following models were adapted to traffic on freeways. For the application in urban areas, the new model T^3 was developed at Bosch by using measurements that were recorded with a test car. The car was equipped with a radar sensor to measure the distance and relative speed of the front car in addition to its own speed and acceleration. Based on these measurements the model was developed using a consequently data driven approach, see [8]. The implemented T^3 model is now used inside ARTIST for urban simulation tasks.

6 Conclusion

Using traffic simulation for the design of traffic management systems has proven to be a powerful tool for the assessment and optimization of traffic control strategies. It offers the opportunity to do this task in a cost-effective and safe way. With this approach, the impact of planned measures can be studied in a reproducible way, even at early phases in the design process. Traffic simulation demonstrably leads to better products.

On the other hand there are still some open issues to take into account. First of all, simulation is not yet common practice in the traffic community. Sometimes one has to convince decision makers on the customer's or the traffic engineer's side to use such an approach at all. But experience has shown that once introduced, traffic simulation is always appreciated in the end.

On the technical side, more research on traffic models is still needed. While car following and traffic flow effect is state of the art in most traffic models, there is still a deficiency concerning lane changing, navigation, complex right-of-way handling, and compliance with traffic signs. Nearly every microscopic traffic model uses components to cover these aspects but most of the behavior is more ad hoc with sometimes plausible effects, rather than being based on data gathered in real traffic. One reason for this is the lack of measurements to calibrate traffic models to the real world. With new traffic sensors to be expected in the cars and on roads in the near future, improvements in measuring can be achieved, which will also allow traffic models to be refined.

References

1. G. Reichart, „MOTIV - A cooperative research programme for the mobility in urban areas". *Proc. 4th World Congress on ITS*, Berlin, Germany, Oct. 21-24, 1997.

2. W. Krautter, D. Manstetten, and T. Schwab, „Traffic simulation with ARTIST". *Proc. IEEE Conf. on ITS*, Boston, Mass., Nov. 9-12, 1997.

3. The MathWorks, Inc, *Simulink: Dynamic System Simulation Software - User's Guide*, 1992.

4. Integrated Systems Inc., *SystemBuild/WS, V2.4 - User's Guide*, 1991.

5. Der Bundesminister für Verkehr, Abteilung Straßenbau: „Technische Lieferbedingungen für Streckenstationen - TLS". *Verkehrsblatt-Sammlung NR. S 1078*, Verkehrsblatt-Verlag, Dortmund, Germany, 1993.

6. D. Manstetten and J. Maichle, „Determination of traffic characteristics using fuzzy logic“. *Proc. 3rd World Congress on ITS*, Orlando, Florida, Oct. 14-18, 1996.

7. D. Manstetten, W. Krautter, and T.Schwab: „Traffic simulation supporting urban control system development“. *Proc. 4th World Congress on ITS*, Berlin, October 21-24, 1997.

8. T. Bleile: „A new microscopic model for car-following behavior in urban traffic“. *Proc. 4th World Congress on ITS*, Berlin, Germany, Oct. 21-24, 1997.

9. R.Wiedemann: *Simulation des Straßenverkehrsflusses*. Schriftenreihe des Instituts für Verkehrswesen, Univ. Karlsruhe, 1974.

10. M. Bando et al., „Dynamical model of traffic congestion and numerical simulation“. *Physical Review E*, Vol. 51, No. 2, 1995.

11. M. Schreckenberg, A. Schadschneider, K. Nagel, and N. Ito: „Discrete stochastic models for traffic flow“. *Physical Review E*, Vol. 51, No. 4, 1995.

12. Q. Yang and H.N. Koutsopoulos: „A microscopic traffic simulator for evaluation of dynamic traffic management systems“. *Transortation. Research C*, Vol. 3, No. 4, 1996.

13. S. Witte: *Simulationsuntersuchungen zum Einfluß von Fahrerverhalten und technischen Abstandsregelsystemen auf den Kolonnenverkehr.* PhD Thesis, Univ. Karlsruhe, 1996.

14. M. Cremer: *Der Verkehrsfluß auf Schnellstraßen*. Springer-Verlag, Berlin, 1979.

Modelling Advanced Transport Telematic Applications with Microscopic Simulators: The Case of AIMSUN2

J. Barceló[1], J. Casas[2], J.L. Ferrer[1], and D. García[2]

[1]Laboratori de Simulació i Investigació Operativa, Departament d'Estadística i Investigació, Operativa, Universitat Politècnica de Catalunya, Pau Gargallo 5, 08028 Barcelona, Spain

[2]TSS-Transport Simulation Systems, Tarragona 110-114, 08015 Barcelona, Spain

The simulation of Advanced Transport Telematic Applications requires specific modelling features, which have not been usually taken into account in the design of microscopic traffic simulation models. This paper discusses the general requirements of some of these applications, and describes how have they been implemented in the microscopic traffic simulator AIMSUN2.

1 Introduction

Microscopic traffic simulators are simulation tools that emulate realistically the flow of vehicles on a road network. The main modelling components of a microscopic traffic simulation model are: an accurate representation of the road network geometry, a detailed modelling of individual vehicles behaviour, and an explicit reproduction of traffic control plans. The primary attention has been paid usually to the proper modelling and calibration of all these model components, namely the car-following, gap acceptance, lane change, and other internal models which along with other modelling parameters accounting for attributes of the physical system entities, allow the microscopic simulation model reproduce flow, speed, occupancies, travel time, average queue lengths, etc. with enough accuracy to consider the model valid.

The advent of the Advanced Transport Telematic Applications made possible by combining the developments in informatics and telecommunications applied to transportation problems, has created new objectives and requirements for micro-simulation models. Quoting from Deliverable D3 of the SMARTEST Project [1]: "The objective of micro-simulation models is essentially, from the model designers point of view, to quantify the benefits of Intelligent Transportation Systems (ITS), primarily Advanced Traveller Information Systems (ATIS) and Advanced Traffic Management Systems (ATMS). Micro-simulation is used for evaluation prior to or in parallel with on-street operation. This covers many objectives such as the study of dynamic traffic control, incident management schemes, real-time route guidance strategies, adaptive intersection signal controls, ramp and mainline metering, etc.

Furthermore some models try to assess the impact and sensitivity of alternative design parameters".

The current trend in the development of Advanced Transport Telematic Applications, either real-time adaptive, or based on other specific approaches, is far from being standardised. It is therefore an exercise of dubious utility to try to integrate them in a fixed way in a microscopic traffic simulator. The relative gain achieved by including any of these, as an in-built function of the microsimulator is limited to simulating, on an easier way, those road networks on which the selected application is operating. However there would be no means of simulating other systems with that microsimulator. This is true whenever we address the problem of simulating adaptive traffic control systems as, for example, SCOOT, SCATS, vehicle actuated, control systems giving priority to public transport, etc., Advanced Traffic Management Systems (using VMS, traffic calming strategies, ramp metering policies, etc), Vehicle Guidance Systems, Public Transport Vehicle Scheduling and Control Systems or applications aimed at estimating the environmental impacts of pollutant emissions, and energy consumes. The main question then is: How can these Advanced Transport Telematic Applications be properly evaluated and tested by simulation?

From a conceptual point of view the operation of these modern systems can be described as follows: for certain applications the road network is suitably equipped with traffic detectors of various technologies (loop detectors, image processing detectors, etc.), with a specific layout depending on the requirements of the control approach. They supply the necessary real-time traffic data (flows, speeds, occupancies, etc) with the required degree of aggregation. These real-time traffic measurements feed the logic of the traffic control or management system which, after suitable processing, makes ad hoc control decisions: e.g. extend the green phase, change to the red phase, apply some traffic calming strategies, etc.. Other applications as, for example vehicle guidance, public transport monitoring systems, or the evaluation of environmental impacts, require the access to vehicle data (position, speed, acceleration, etc.), to emulate the up-link messages in vehicle guidance applications, the vehicle tracking for the public transport monitoring or fleet management systems, or simply to provide the required data for certain fuel consumption or pollutant emissions models. To evaluate and test any of these systems a microsimulator must be capable of incorporating in the model the corresponding traffic devices as objects: i.e. detectors, traffic lights, VMS, etc. It must also emulate their functions: provide the specific traffic measurements at the required time intervals, increase the phase timing in a given amount of time, implement a traffic calming strategy (slow down the speed on a road section, recommend an alternative route, etc). How can such evaluations be done by simulation without explicit in-built modelling of the specific Advanced Telematic Application?

2 The GETRAM/AIMSUN2 Microsimulator

2.1 GETRAM

GETRAM (Generic Environment for Traffic Analysis and Modelling), [2, 3], is a simulation environment comprising a traffic network graphical editor (TEDI), a microscopic traffic simulator (AIMSUN2), a network data base, a module for storing results and an Application Programming Interface to aid interfacing to assignment models and other simulation models, as for example the macroscopic traffic assignment EMME/2 system, [4]. The functional structure of the systems is depicted in Fig. 1.

2.2 AIMSUN2 Microscopic Simulator

AIMSUN2 (Advanced Interactive Microscopic Simulator for Urban and Non-Urban Networks), [5], is a microscopic traffic simulator that can deal with different traffic networks: urban networks, freeways, highways, ring roads, arterial and any combination of them. AIMSUN2 simulates traffic flows either based on input traffic flows and turning proportions, or on O-D matrices and route selection models. In the former, vehicles are distributed stochastically around the network, whereas in the latter vehicles are assigned to specific routes from the start of their journey to their destination. Different types of traffic control can be modelled in AIMSUN2: traffic signals, junctions without traffic signals (give way or stop signs) and ramp metering. Vehicle behaviour models (car following, lane change, gap acceptance, etc.) are functions of several parameters that allow modelling of different types of vehicles: cars, buses, trucks, etc. They can be classified into groups, and reserved lanes for given groups can also be taken into account.

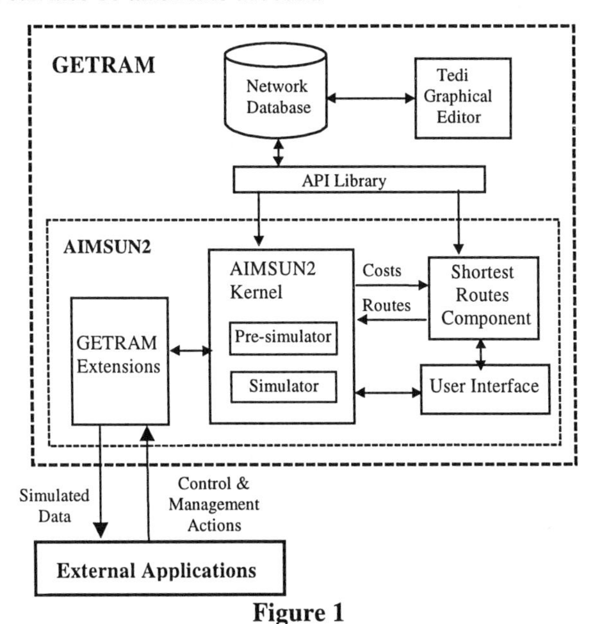

Figure 1

The logic of the simulation process in AIMSUN2 is illustrated in the diagram of Fig. 2. It can be considered as a hybrid simulation process combining an event scheduling approach with an activity scanning. At each time interval (simulation step), the simulation cycle updates the unconditional events scheduling list, that is events like traffic light changes which not depend on the termination of other activities. The "Update Control" box in the flow chart represents this step. After this updating process starts a set of nested loops updating the states of the entities (road sections and junctions) and vehicles in the model. Once the last entity has been updated the simulators performs the remaining operations: input new vehicles, collect new data, etc.

Depending on the type of simulation new vehicles are placed into the network according to flow generation procedures (headway distributions for example) at input sections, or using time sliced O-D matrices and explicit route selection, as shown in the diagram. The simulation process includes in this case an initial computation of routes going from every section to every destination according to link cost criteria specified by the user. A shortest route component calculates periodically the new shortest routes according to the new travel times provided by the simulator, and a route selection model assigns the vehicles to these routes during the current time interval. Vehicles hold the assigned route from origins to destinations at least they had been identified as "guided" at generation time, then they can dynamically change the route en route as required for simulating vehicle guidance and vehicle information systems.

The car following model implemented in AIMSUN2 is based on the Gipps model [6], and can be considered as an ad hoc evolution of this empirical model in which the model parameters are not global but determined by the influence of local parameters depending on the "type of driver" (limit speed acceptance of the vehicle), the geometry of the section (speed limit on the section, speed limits on turnings, etc.), the influence of vehicles on adjacent lanes, etc. The lane change model in AIMSUN2 can also be considered as an further evolution of the seminal Gipps lane change model [7]. Lane change is modelled as a decision process analysing the necessity of the lane change (as in the case of turning manoeuvres determined by the route), the desirability of the lane change (as for example to reach the desired speed when the leader vehicle is slower), and the feasibility conditions for the lane change that are also local, depending on the location of the vehicle on the road network.

AIMSUN2 can also simulate any kind of measurable traffic detector: counts, occupancy and speed. AIMSUN2 has a user-friendly interface through which the user can define the simulation experiment. It also provides a picture of the network and an animated representation of the vehicles in it. The user has an overview of what is happening in the network that aids performance analysis. Through the interface, the user may access any information in the model and define traffic incidents before or during the simulation run.

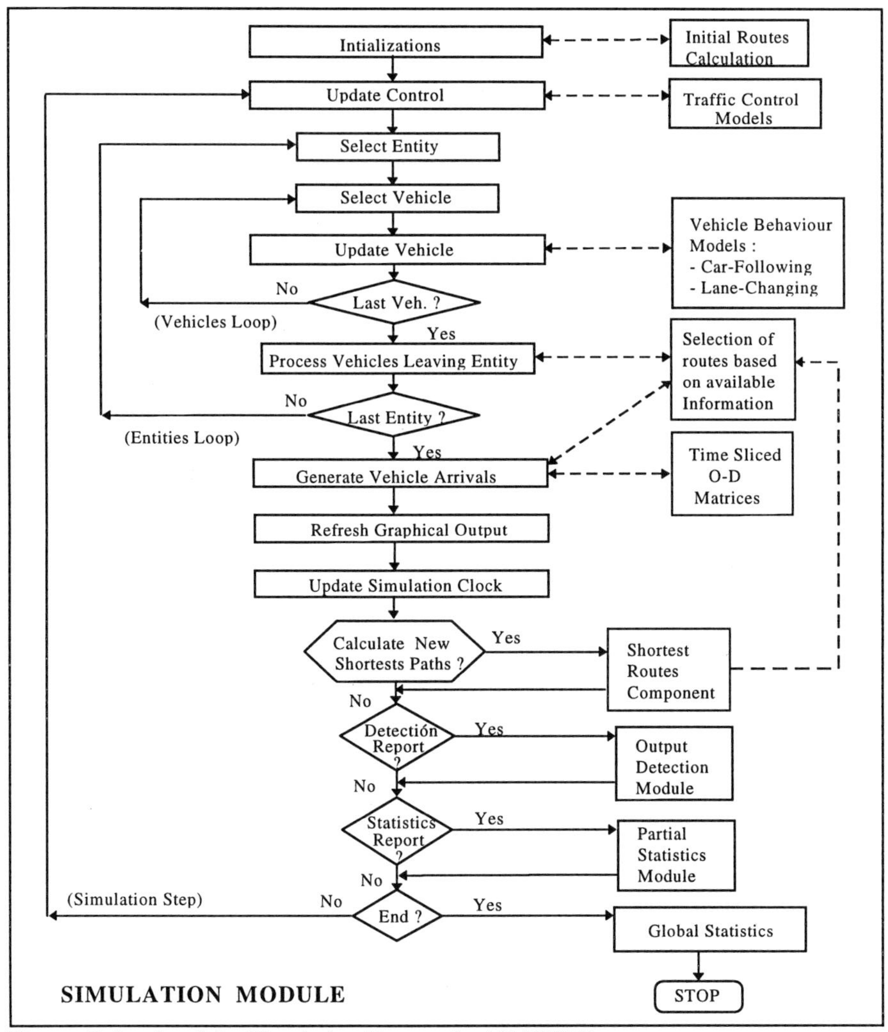

Figure 2: Logic of the Simulation Process in AIMSUN2.

2.3 TEDI Network Editor

TEDI is a graphical editor for traffic networks. It has been designed with the aim of making the process of network data entry and model building user-friendly. Its main function is the construction of traffic models with which to feed traffic simulators like AIMSUN2. To facilitate this task the editor accepts as a background a graphical description of the network area, so sections and nodes can be built subsequently into the foreground. The editor supports both *urban* and *interurban* roads, which means that the level of detail covers elements such as side lanes, entrance and exit ramps, intersections, traffic lights and ramp metering. TEDI has an interface to the EMME/2 DATA BANK, providing the means to complement a macroscopic analysis effortlessly with a microscopic one using the same traffic data (i.e. O/D matrices). The user can define a hierarchical tree of views so that a traffic model can be restricted to one of these views. The editor is designed for the level of detail required by regional, intermediate and local areas, with increased modelling detail. A library of high-level, object-based application programming functions, named TDFunctions, assists the development of interactive external applications, and, in general to access any data. The TDFunctions enable objects in the network to be read and manipulated or can restrict the view to a sub-area. Results storage and control plans are also accessed with these functions.

3 GETRAM/AIMSUN2 Extensions

To cope with the requirements of simulating Advanced Transport Telematic Applications specific extensions to GETRAM/AIMSUN2 have been developed. These extensions fall into three categories:

3.1 Adaptive Traffic Control, Traffic Management Systems and Incident Management Systems .

3.2 Vehicle Guidance, Fuel Consumption and Emissions.

3.3 Public Vehicle Scheduling and Control Systems.

The approach taken in GETRAM/AIMSUN2 consists of considering the Advanced Telematic Application to be tested as an EXTERNAL APPLICATION that can communicate with GETRAM/AIMSUN2. An ad hoc version of AIMSUN2 including a set of DLL has been developed for this purpose. This library gives AIMSUN2 the ability to communicate with almost any of the above-mentioned external applications.

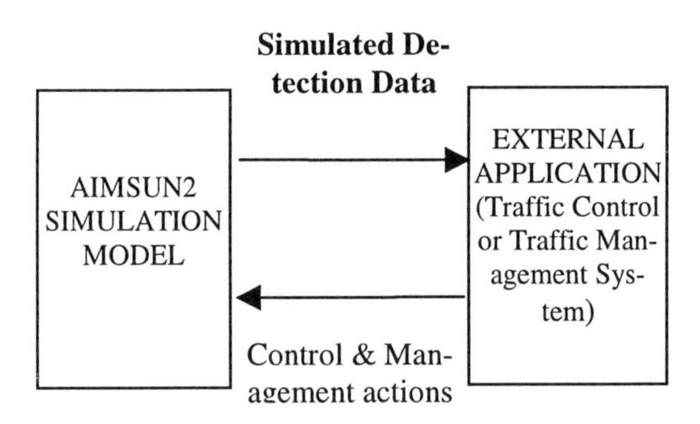

Figure 3

Using the TEDI & AIMSUN2 functions the detector, VMS and traffic lights can be modelled and their attributes defined. The process of information exchange between AIMSUN2 and the external application is shown in Fig. 3. The AIMSUN2 model of the road network emulates the detection process providing the external application with therequired "*Simulation Detection Data*". The EXTERNAL APPLICATION (user provided) decides which control and/or management actions have to be applied on the road network and sends the corresponding information to the simulation model which then emulates their operation through the corresponding model components such as traffic lights, VMS, etc. Another set of DLL function enables the user to access the information on each vehicle state (position, speed, acceleration, etc.) at each simulation cycle.

3.1 Simulating Management and Control Actions

Three main types of actions, as a result of the actuation of EXTERNAL APPLICATION on the simulation, are taken into account:

1) Actuate control of traffic lights and ramp metering.

2) Actuate control of ramp metering.

3) Supply information to the driver using Variable Message Signs (VMS).

3.1.1 Traffic Control Actions

AIMSUN2 takes into account two types of traffic control: traffic lights and ramp metering. The first is considered for urban type intersection nodes, while the second one is for controlling freeway entrance ramps.

For the intersection control, a phase-based approach is applied in which the cycle of the junction is divided into phases where each one has a particular set of signal groups with right of way (a signal group is considered as one traffic light). During the simulation of a scenario, AIMSUN2 executes a fixed control plan taking into account the phase modelling for each junction. An EXTERNAL APPLICATION can modify this execution by means of different actions. The available actions are to:

1. Change the duration of each phase: The EXTERNAL APPLICATION can increase or decrease the duration, but the control plan structure is not modified.
2. Disable the fixed control plan structure: The EXTERNAL APPLICATION disables the structure of the control plan and completely controls the phase changing.
3. Change the current phase: The EXTERNAL APPLICATION can change the current phase to another. If the fixed control plan (timings) of the junction have not been disabled, AIMSUN2 programmes the next changing of phase taking into account the duration of the phases. Otherwise AIMSUN2 holds the new phase until the EXTERNAL APPLICATION changes it to another.

Examples of functions of the DLL library relative to control junctions are:

- Read Number of junctions*: Reads the number of junctions present on the road network.*
- Read the Identifier of a junction: *Reads the identifier of junction elem-th present on the road network.*

- Read the Name of a junction: *Reads the name of junction identified by idJunction.*
- Read the number of Signal Groups of a junction: *Reads the number of signal groups defined in junction identified by idJunction.*
- Read the Number of Phases of a junction: *Reads the total number of phases of a junction in the current control plan.*
- Read Time Duration of a phase of a junction: *Reads the maximum, minimum and current duration of a in junction during the current control plan.*
- Read the Current Phase of a junction: *Reads the current phase of a junction, even when the junction has disabled the fixed control plan.*
- Disables the fixed control plan of a junction: *Disables the fixed control plan of a junction, so the phase changing is completely controlled by the EXTERNAL APPLICATION (i.e. an adaptive traffic control system).*
- Change of Phase: *Changes to idPhase in junction identified by idJunction.*
- Change the Current Duration of Phase: *Changes the duration of idPhase in junction identified by idJunction in the current control plan.*
- Change the State of a Signal Group: *Changes the state of signal group idSigGr in junction identified by idJunction.*

3.1.2 Ramp Metering

AIMSUN2 also incorporates ramp-metering control. This type of control is used to limit the input flow to certain roads or freeways in order to maintain certain smooth traffic conditions. The objective is to ensure that entrance demand never surpasses the capacity of the main road. AIMSUN2 considers three types of ramp metering depending on the implementation and the parameters that characterise it:

1. Green time metering, with parameters green time and cycle time. It is modelled as a traffic light.
2. Flow metering, with parameters platoon length and flow (veh/h). The meter is automatically regulated in order to permit the entrance of a certain maximum number of vehicles per hour.
3. Delay metering, with parameters mean delay time and its standard deviation. It is used to model the stopped vehicles due to some control facility, such as a toll or a customs checkpoint.

The EXTERNAL APPLICATION can modify this modelling by different actions. It can:

1. Change the parameters of a metering, the EXTERNAL APPLICATION can dynamically modify the parameters that define a ramp metering.

2. Disable the control structure: EXTERNAL APPLICATION disable the structure of the ramp metering and completely controls the state changing.

3. Change the state of a metering: The EXTERNAL APPLICATION can change the current state to another. If the metering has not disabled the control, AIMSUN2 programmes the next changing of state taking into account the parameters, which define the control. Otherwise AIMSUN2 holds the new state until the EXTERNAL APPLICATION changes it to another.

Examples of functions of the DLL relative to ramp metering are:

- Read the Section Identifier of a metering: *Reads the section identifier that contains the metering elem-th present on the road network.*

- Read the Type of metering: *Reads the type of metering present in an section. (The type of a metering can be 0: None; 1: GREEN METERING; 2:FLOW METERING; 3: DELAY METERING.)*

- Read the Control Parameters of a Green Metering: *Reads the parameters of a green metering that are defined in the current control.*

- Change the Control Parameters of a Green Metering: *Changes the parameters of a green metering that are defined in the current control.*

- Read the Control Parameters of a Flow Metering: *Reads the parameters of a flow metering defined in the current control.*

- Change the Control Parameters of a Flow Metering: *Changes the parameters of a flow metering that are defined in the current control.*

- Read the Control Parameters of a Delay Metering: *Reads the parameters of a delay metering that are defined in the current control.*

- Change the Control Parameters of a Delay Metering: *Changes the parameters of a delay metering that are defined in the current control.*

- Disable the fixed control plan of a metering: *Disables the fixed control plan of a metering, so the state changing is completely controlled by the EXTERNAL APPLICATION.*

3.1.3 Variable Message Signs

Providing information to drivers is a possible action of a Traffic Management System on a road network equipped with Variable Message Sign infrastructure. Messages may inform drivers about the presence of incidents, congestion ahead or suggest alternative routes. AIMSUN2 takes into account the modelling of Variable Message Sign (VMS) as defined in TEDI by means of a dialogue including the VMS name or identification code, its position in the section, the activated message, if any, the list of feasible messages for this VMS, and the list of all Actions available for this network associated to the messages.

Messages in a VMS from its message list may be activated in two different ways: directly through the user interface or by an external application through the communication interface. In both cases, it will cause the message to be displayed as Activated Message and the Actions associated with it to be implemented. Each message has a list of Actions associated with it which appear in the list box named 'Mess Actions' in the VMS Information Window. The list box named 'Actions' contains all actions available for this network. An Action represents the expected impact a message has on driver's behaviour. Examples of Actions are: modifications of the speed limit, modification of the input flow, modifications of the turning proportions.

When simulating with the Route Based option actions can also imply a re-routing, that is the possibility of altering the vehicle's path. This effect is accomplished by defining the next turn and/or defining a new destination. The re-routing effect is defined by the following for each modality independently:

- **Compliance level** (δ): This parameter gives the compliance level of the action. If $\delta=1$then it causes the re-routing to be followed by all vehicles (i.e. it is obligatory). If $\delta=0$ then the re-routing action will be followed depending on the driver behavioural parameter (a local parameter of it), i.e. it is an information only. When $0<\delta<1$, δ gives the level of acceptance, e.g. it is advice.
- **Modify the next turning**: Change, which is the next turning that the vehicle must follow. This action is defined taking into account the destination.
- **Modify destination**: When a vehicle enters into a section affected by an action, the simulator changes its destination.

Functions relative to VMS of the DLL library are:

- Read the Identifier of a VMS*: Reads the section identifier that contains the VMS elem-th present on the road network.*
- Read the message of a VMS: *Reads the text of a message elem-th in a VMS.*
- Read the Current Active Message: *Reads the current active message in a VMS.*
- Active a Message in a VMS: *Actives in PanelName the Message. AIMSUN2 executes the actions associated in Message.*

3.1.4 Detector Measurements

Detection output data is produced by AIMSUN2 periodically, provided that there are detectors defined in the network and the Detection Function of the simulator is activated. Currently there are two main types of Detection implemented: Common Detection Model and EXTERNAL APPLICATION type Detection Model.

In the Common Detection Model, the data produced depends on the measuring capabilities of the detectors. There is a data line for each detector, which contains the detector identifier and the list of measures gathered. They may be Count (number of vehicles per interval), Occupancy (percentage of time the vehicle is on the detector),

Speed (mean speed for vehicles crossing the detector) and Presence (if a vehicle has been on the detector, it is set to 1). These data are stored in ASCII files.

In the EXTERNAL APPLICATION type Detection Model, the measures are given at every simulation step or aggregated each detection interval. The gathered measures are: Counts (Number of vehicles), Speeds (Mean speed for vehicles crossing the detector), Occupancies (percentage of time the detector is activated), and Presence (whether a vehicle is over the detector or not). The EXTERNAL APPLICATION can undertake the following actions with detectors: retrieve the number of detectors in the network, retrieve the name of each detector, retrieve the detection interval, retrieve the detector measures gathered in each simulation step, retrieve the aggregated detector measures.

3.2 Simulating Incidents and Incident Management

The diagram in Fig. 4 schematizes the methodological procedure proposed for the simulation of incident detection and management based on the EXTERNAL APPLICATIONS. The procedure is based on a microscopic simulation model of a site that emulates traffic conditions on the site, and generates traffic data: flows, occupancies, speeds, (travel times when required), at the sampling rate requested by the external applications (for example 30 seconds is an standard request for most automatic incident detection algorithms, [8]), with the format proper of the technology used at the site. These traffic data feed the Incident Warning, Incident Detection and Traffic Management Modules implementing the corresponding External Applications.

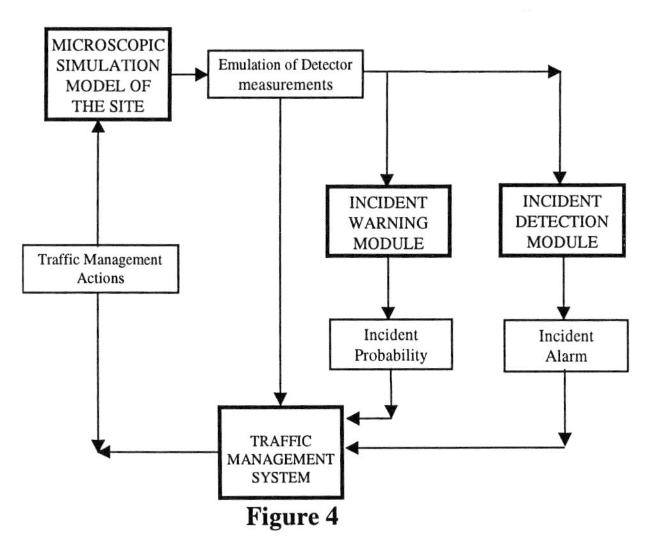

Figure 4

The Incident Warning applications estimate an incident probability [9] that is sent as a warning to the Traffic Management System that may take it into account. The Simulation model dialogues with the Management System, as an external application, in the way described above. Once the Incident Detection Module detects an incident, it generates an incident alarm, which is sent to the Traffic Management System. The management decisions are communicated to the simulation model through the proper dialogue as described above.

3.3 Simulation of Vehicle Guidance

Vehicle Guidance can also be modelled as an External Application that can be properly simulated with AIMSUN2 by means of an exchange of information using the suitable DLL functions. Essentially the simulation of a Vehicle Guidance system, [10], consists of an exchange of information between the equipped car and the Traffic Information Centre. The on board equipment collects information on the vehicle's position, travel time, speed, experienced delay, number of stops, etc. which is sent to the Traffic Information Centre, [11-13], by means of the telecommunication technology on which the system is based, as for example, beacons or GSM, or both.

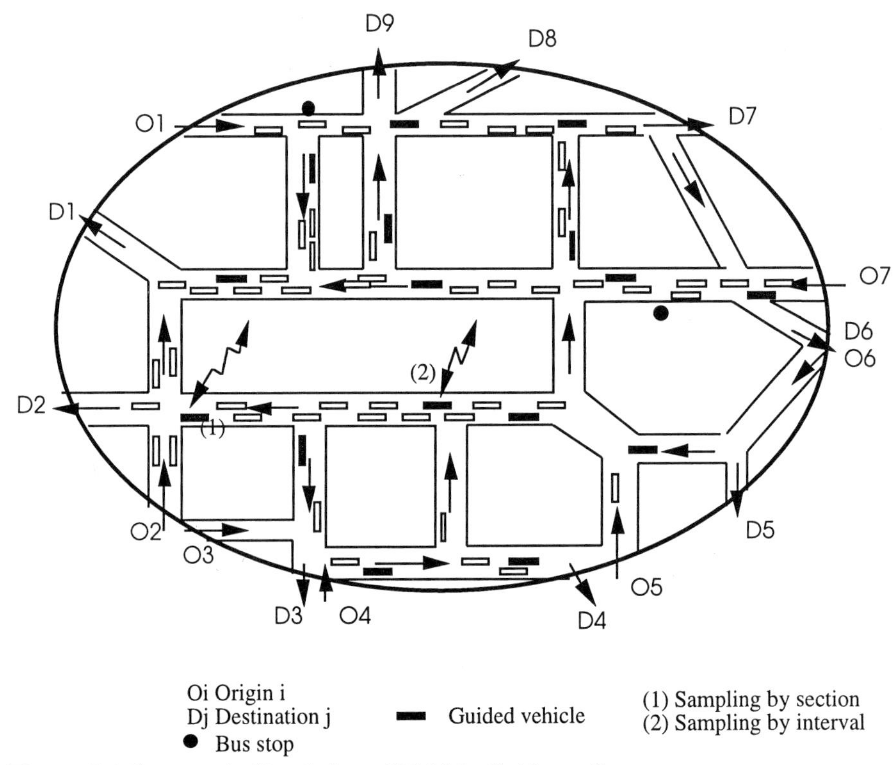

Figure 5: Microscopic Simulation of Vehicle Guidance Systems.

The vehicle data collection can be modelled as follows: each equipped vehicle sends the information to the system when passing through certain points on the network. In the network description the definition of these Data Collection Points (DCP) have been included working like detectors. Each time the position of a guided vehicle is updated it is checked whether or not it has passed through a DCP. If so, a message to the information centre is sent and the vehicle information is updated.

Alternative data collection procedures can be implemented for the simulation, as for example assuming that there is a set of fixed DCP which correspond to the end of each section, or that DCP are variable and their position depends on the behaviour of each guided vehicle, for instance, if it takes more than certain time for a vehicle to cross a section or if it has to stop during a section journey. Examples of floating car data collection sampling procedures, as illustrated in Fig. 5, are: section based (a vehicle sends a message whenever it reaches the end of any section in the network); time based (a vehicle sends a message every certain time interval whose length can be selected by the user as a simulation input parameter); dual mode sampling (time and section based, if a vehicle takes more than one time interval to travel one section, a message is sent for each time interval); speed and section basis: a vehicle sends a message at the end of each section. Besides, if during the trip along a section the vehicle speed falls below a minimum value (which is set by the user as a simulation input parameter), a message is sent every time interval until the end of section is reached.

Guided vehicles, which are a percentage of the total, are identified at generation time and the data collection process is initialized. For each guided vehicle the corresponding DLL functions collect the information on the vehicle data. Additionally, each time a guided vehicle is updated, it is checked wether or not it is time to transmit data. If so, the following information is updated by means of the ad hoc DLL: Number of messages sent, Number of data blocks transmitted, Sections and distance travelled since last message, Time for the next transmission. In a similar way every time a guided vehicle reaches the end of a section, the following information is updated: Number of messages sent, Number of data bolcks transmitted, Distance travelled since last message, Time since last transmission.

The simulation can thus provide the following information [14]: trafic volumes (guided and unguided); mean vehicles speed (km/h); number of stops per vehicle/kilometer; communications overhead in total number of messages between vehicles and information center, per time unit, and total number of blocks transmitted per time unit; mean time between messages per vehicle; average distance travelled between messages per vehicle and average number of messages per section and vehicle. The first three items estimate the level of network congestion in the simulation experiment. The next two estimate the communications requirements for the system, and the last three are measures about the quality of the overall information received by the centre. To illustrate these simulation results let us summarily describe a typical simulation experiment of a guidance system. The parameters defining the experiment are the following: **MNV**, mean number of vehicles in the network; **MNG**, mean number of guided vehicles in the network; **NTR**, number of trips / hour; trips from origins to destinations in the modeled network; **SPD**, mean journey speed in kilometers/hour; **NST**, number of stops per vehicle per kilometer traveled. The results on communications requirements produced as output by the AIMSUN2 simulation are represented by the following variables: **MpS**, number of messages (transmissions) per second sent from guided vehicles to the information center; **BpS**, number of data blocks transmitted per second. For instance, a trip message is composed by one data block for the header and one data block for each section traveled. Taking into account the type of information of each section of the messages, the typical lengh of a data block could be considered as 16 bytes. AIMSUN2 output provides also an estimate of the quality of the overall data transmitted from the equipped vehicles to the information centre by means of the following variables: **TbT**, mean time between transmissions per vehicle, which is a measure of how often the center has information from a given vehicle; **DbT**, mean distance travelled between transmissions per vehicle, this measure together with the **TbT**, gives an idea about the frequency of information updating for each vehicle; **CpS**, mean number of messages transmitted per section per vehicle. This provides measures of the information quality for each section.

Thus, for example, for a simulation experiment done in the SOCRATES project [12, 14], the traffic conditions where characterised by the following values:

MNV	MNG	NTR	SPD	NST
10000 veh	500 veh	54000 trips	30 km/h	6.2 stops

Therefore, the mean number of guided vehicles that the system was controlling was 500 vehicles from a total of 10000 (that is the 5%). The mean speed on the network was 30 km/h and a vehicle had to stop on average about 6 times every kilometre travelled.

The communications requirements obtained in that simulation experiment for the different sampling procedures are summarized in the following Tab.:

	section	time 30	time 60	dual 30	dual 45	dual 60	speed 30	speed 45	speed 60
MpS	10.3	16.3	8.0	17.5	13.7	12.0	19.7	17.6	16.7
BpS	20.6	42.2	25.3	42.6	31.2	26.1	49.2	42.9	40.2

The quality of the data received by the traffic information centre can be represented by the following measures:

	section	time 30	time 60	dual 30	dual 45	dual 60	speed 30	speed 45	speed 60
TbT	44.5”	30.0”	60.0”	21.3”	27.0”	30.8”	18.9”	21.1”	22.2”
DbT	299 m	185 m	366 m	171 m	217 m	248 m	152 m	170 m	180 m
CpS	1	1.57	0.77	1.76	1.38	1.21	1.98	1.77	1.68

From this experiment, it appears that the best sampling procedures are those based on dual mode sampling on time basis. For not to fall in communication overheads, a time interval of 60 sec. or more is preferable.

The Traffic Information Centre collects the individual information from the equipped vehicles and, after a suitable processing produces the guidance information which is transmitted to the guided vehicles. The broadcasting of this information and the guided vehicle reactions can be simulated using the DLL in a similar way as the simulation of the VMS, based on the capability of the simulator to dynamically re-route the guided vehicles en-route.

4 The Parallelization of AIMSUN2

To conclude this description of improved modelling features in the microscopic simulator AIMSUN2 we will make some comments on what could be expected from the parallelization of the simulator. A complete description can be found elsewhere, [15-17]. Some reasons to parallelize a microscopic traffic simulator could be the following: the situation of the current practices in traffic management can be summarized saying that, out of a limited traffic control practice, traffic management is currently based on manual procedures, relying on the experience of human operators, and off-line computing practices. There is a main reason for that, numerical algorithms, based either on optimization or simulation approaches, to deal with time-varying traffic flows in real size traffic networks have very high computing requirements to be processed sequentially on the currently available computing platforms, the resort to parallel computing is a way for achieving the required performance for real-time applications.

The encouraging results obtained with the parallel version of AIMSUN2, reported in [17], based on a standard library of threads available in the current version of SUN Solaris 2.4, and a shared memory computing platform, where average speed ups of up to 3.5 times have been achieved, show that a quasi-real time operation of the systems using microscopic simulation for management purposes is feasible. In the computational experiments only the basic version of AIMSUN2 has been parallelized, certainly the inclusion of routing information and its use in the microsimulation will impose additional computational burden that must be investigated. On the other hand the parallelization studied depends, obviously, on the structure of AIMSUN2 and therefore the results cannot be extrapolated to other microscopic simulator with different internal structures. However, we believe that our results show that the parallelization of the microscopic simulators, on the currently available computer platforms, opens the door to simulation analysis of medium to large networks, and not only small networks as microscopic approaches had been restricted so far, and to the use of simulation as decision support tool in the context of Traffic Management applications.

5 Conclusions

The paper has described the requirements to simulate Advanced Transport Applications and the conceptual approaches to implement them in microscopic traffic simulators. The description has been completed showing how these approaches have been implemented in the case of the AIMSUN2 microscopic traffic simulator. These implementations have been mainly done through the participation in projects of the ATT Programme of the European Union. References to these implementations can be found in the already referenced reports of SMARTEST (simulation of VMS, ramp metering, etc.) [1], SOCRATES (Vehicle Guidance) [11, 12], CLAIRE SAVE (Adaptive Traffic Control and Environmental impacts) [18], CAPITALS (Traffic Management) [19], IN-RESPONSE (Incident Management) [20], and PETRI, [16].

References

1. SMARTEST Project Deliverable D3, August 1997, European Commission, 4th Framework Programme, Transport RTD Programme, Contract Nº: RO-97-SC.1059.

2. R. Grau and J. Barceló. *The design of GETRAM: A Generic Environment for Traffic Analysis and Modeling*. Research Report DR 93/02. Departamento de Estadística e Investigación Operativa. Universidad Politécnica de Cataluña. (1993).

3. J. Barceló, J.L. Ferrer and R. Grau. *AIMSUN2 and the GETRAM Simulation Environment.* Technical Report. Departamento de Estadística e Investigación Operativa. Universidad Politécnica de Cataluña (1994).

4. INRO Consultants, *EMME/2 User's Manual.* Software Release 8.0 (1996).

5. J. Barceló and J.L. Ferrer. *AIMSUN2: Advanced Interactive Microscopic Simulator for Urban Networks. User´s Manual.* Departamento de Estadística e Investigación Operativa. Facultad de Informática. Universidad Politécnica de Cataluña (1997).

6. P.G. Gipps, *A Behavioural Car-following Model for Computer Simulation*, Transpn. Res. 15B, pp 105-111 (1981).

7. P.G. Gipps, A Model for the Structure of Lane-Changing Decisions, Transpn. Res. 20B, pp.403-414 (1986).

8. Y.J. Stephanedes, A.P. Chassiakos and P.G. Michalopoulos, Comparative Performance Evaluation of Incident Detection Algorithms, Transportation Research Record 1360 (1996).

9. I.R. Wilmink and L.H. Immers, *Deriving Incident Management Measures using Incident Probability Models and Simulation*, TNO Research Report 95/NV/172, The Netherlands (1995).

10. D. Jeffery. Route guidance and In-vehicle Information Systems. In: *Information Technology. Applications in Transport*, P. Bonsall and M. Bell (eds), VNU Science Press, Utrecht, pp. 319-351 (1987).

11. SOCRATES, 1990, DRIVE I Project V1007, Commission of the European Communities, Report on WP1.2.3, *'Floating Car Data'*, Prepared by Hoffmann Leiter, Universitat Politécnica de Catalunya, and BASt, responsibles G.Hoffmann and J.Barceló.

12. SOCRATES KERNEL 1992, DRIVE II Project V2013, Commission of the European Communities, Report SCKN/UPC 03.43.92, *Equipped Vehicle Fleet Requirements for Monitoring Network Conditions in a Dynamic Route Guidance System* Prepared by Universitat Politécnica de Catalunya, responsible J.Barceló.

13. I. Catling, "SOCRATES", in: *Advanced Technology for Road Transport*, ed. By Ian Catling, Artech House, London (1994).

14. J. Barceló, J.L Ferrer, and R. Martín, *Simulation Assisted design and Assessment of Vehicle Guidance Systems*, Accepted for publication in International Transactions on Operations Research, (1998).

15. J. Barceló, J.L. Ferrer, D. García, M. Florian and E. Le Saux, The Parallelization of AIMSUN2 Microscopic Simulator for ITS Applications, Proceedings of the 3rd. World Congress on Intelligent Transport Systems, Orlando (1996).

16. J. Barceló, J.Casas, E.Codina, A.Fernández, J.L.Ferrer, D. García and R.Grau, PETRI: A Parallel Environment for a Real-Time Traffic Management and Information System, Proceedings of the 3rd. World Congress on Intelligent Transport Systems, Orlando, (1996).

17. J. Barceló, J.L. Ferrer, D. García, M. Florian and E. Le Saux, "Parallelization of Microscopic Traffic Simulation for ATT Systems Analysis", in: *Equilibrium and Advanced Transport Modelling*, ed. By P. Marcotte and S. Nguyen, Kluwer (1998).

18. CAPITALS, (1998), "Workpackage 5.2 Evaluation Report", EU DGXIII, Project TR 1007.

19. CLAIRE SAVE, 1996, K. Fox, H. Kirby, G. Scemama and A. Walker, "Avoiding High Fuel Consumption in Congested Cities", Project Final Report, Institute for Transport Studies, University of Leeds.

20. IN-RESPONSE, (1996, 1997) Deliverables 4.3, 5.1 and 5.2, EU DGXIII, Project TR1030.

IV. Environmental Effects

Modelling of Regional and Local Air Pollution Based on Dynamical Simulation of Traffic

W. Brücher[1], M.J. Kerschgens[1], C. Kessler[2], and A. Ebel[2]

[1]Institut für Geophysik und Meteorologie, Bereich Meteorologie, Universität zu Köln, Kerpener Str. 13, 50937 Köln, Germany

[2]Institut für Geophysik und Meteorologie, Projekt EURAD, Universität zu Köln, Aachener Str. 201-209, 50931 Köln, Germany

In the Wuppertal area, Germany, air pollution due to the emission of vehicles is calculated for two summer smog episodes in 1996 and 1997 using a coupled system of atmospheric models that are applicable to different scales. The model system CARLOS (Chemistry and Atmospheric transport in Regional and Local Scale) contains complex subsystems for simulating the atmospheric transfer and the chemistry of gaseous pollutants. The system is capable of reproducing the basic structures of measured data. A higher spatial resolution of the simulations enhances the details of the modelled concentration and wind fields.

1 Introduction

The simulation of air pollution in urban and industrialised areas can only be achieved with spatially high resolving models of the atmospheric chemistry and transfer. This also requires emission inventories which should be highly resolved and detailed in space and time. In this study special emphasis is placed on the contribution of traffic generated emission to the overall air pollution in an urban area. Nested models are used in order to provide realistic time dependant background concentrations at all grid points. By this method local effects are modelled directly and the regional scale influence is simultaneously taken into account via initial and boundary conditions.

In the context of the "Northrhine-Westfalian Research Cooperative Traffic Simulation and Environmental Impact Program" (FVU) the modelling of the concentration of air pollution by traffic emissions is situated last in a row of simulations of different disciplines. In particular, the simulation using CARLOS is designed to analyse and predict the influence of the traffic on the amount of air pollution. Fig. 1 illustrates the principle dependencies and interactions of this project (BEAR) with partners of the FVU project and links to other organisations, which contribute to the necessary flow of input data.

At this stage of development CARLOS is used for the simulation of episodes. Therefore its dynamical part is nested into a global atmospheric circulation model. For this purpose forecasts of the ECMWF model (European Center for Medium range Weather Forecast) are used.

Traffic, industry, household and biogenic sources have to be taken into account as well as background concentrations. For urban scale investigations special emphasis has to be placed upon the simulation of traffic and the resulting emission. In this study the traffic generated emissions are taken from a data set of the Northrhine-Westfalian Environmental Department (LUA) with a resolution of 1-km or from the highly resolved dynamic data of the Cellular Automata (CA) calculated at the Center of Parallel Computing of the University of Cologne (ZPR), depending on the resolution of the model domain.

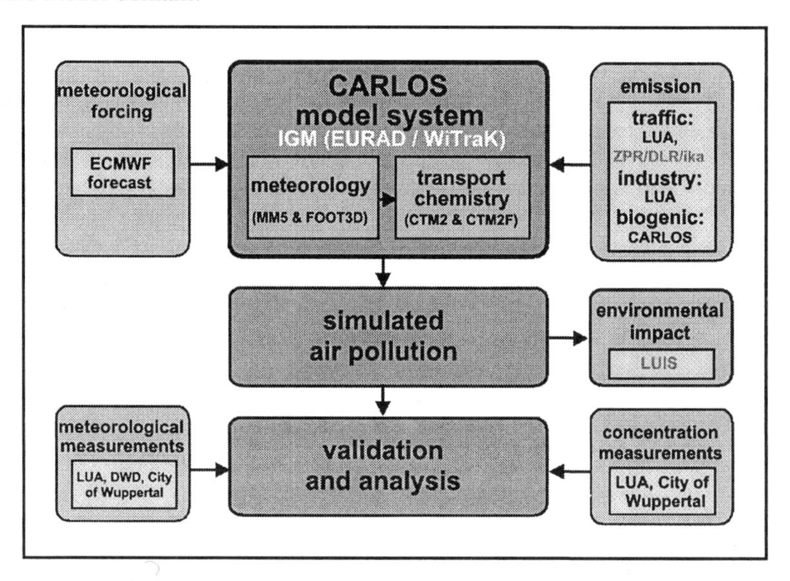

Figure 1: Flowchart of project BAER (**B**elastung der **A**tmosphäre durch verkehrsbedingte **E**missionen und **R**eduzierungsstrategien – Atmospheric pollution by traffic generated emission and strategies for reduction).

By this strategy changes in air pollution due to planned traffic scenarios can be estimated. However, before a prediction of the effects of variable traffic scenarios can be made the present situation needs to be analysed and the models needs to be validated with this analysis. For this purpose multiple nested simulations from the European scale down to the area of Wuppertal have been calculated and validated with measurements. In this article simulations of episodes in summer 1996 and 1997 are presented.

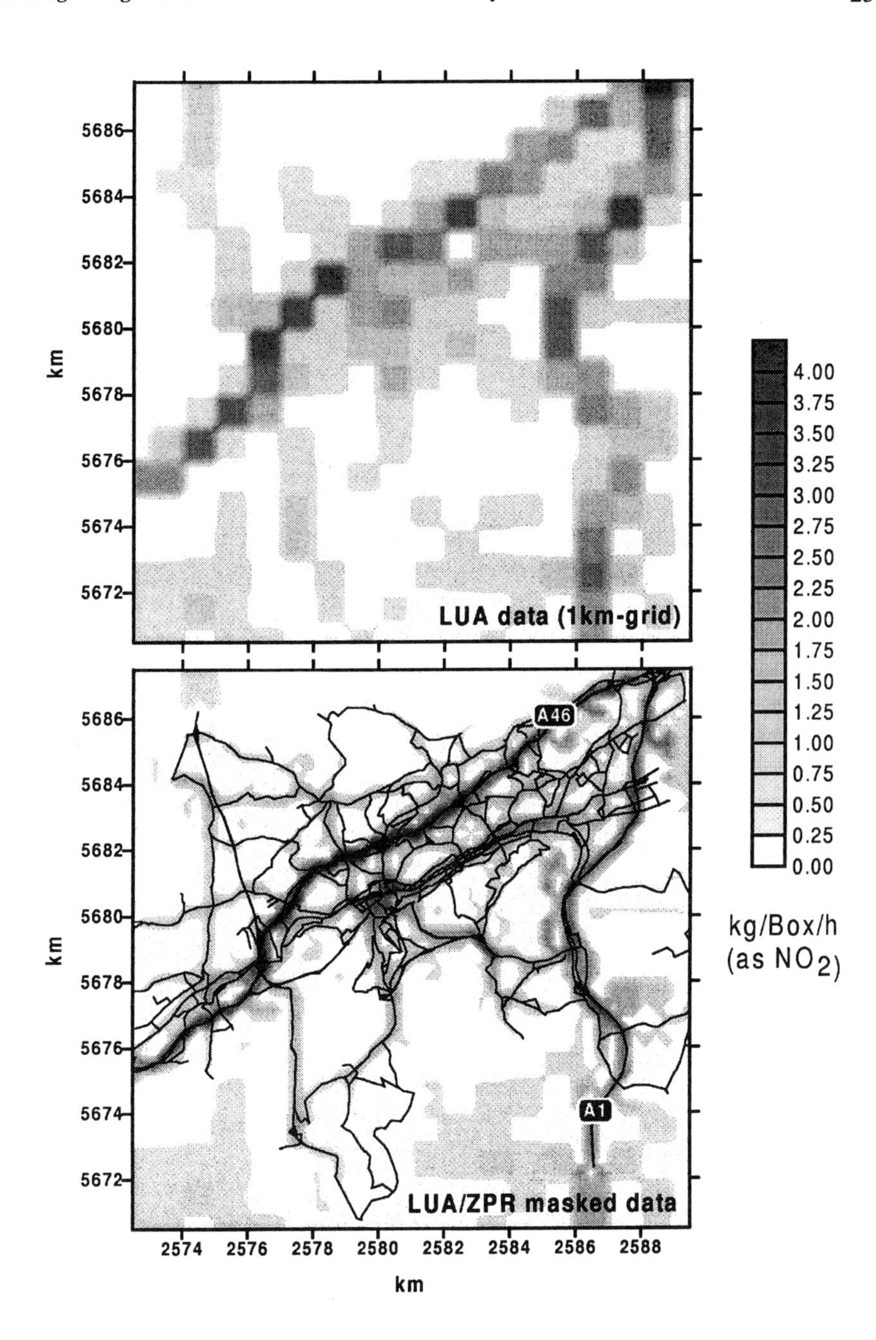

Figure 5: Comparison of NO_X-emissions for the episode in 1997 in the innermost model domain (hourly mean 5-6 UTC). Top: LUA-data, 1km-grid for the large scale simulations. Bottom: ZPR/DLR-data with 333m-resolution combined with LUA-data.

Fig. 6 and Fig. 7 give examples of the concentrations of NO_X at 6 UTC for different resolutions of the model grid. Fig. 6 shows the background concentration modelled by the EURAD system which was taken as the boundary condition for the fine scale CTM2F simulation. At the resolution of 1 km, high concentrations due to the emissions from the highways can already be clearly identified. Enhanced concentrations of NO_X can predominantly be observed where highways are situated in regions of weak wind and weak turbulence. This is the case in valley areas of Wuppertal (W) and Hagen (HA). A more detailed structure can be seen in the pictures of the finest grid. Due to the absence of long distance traffic in the CA simulation, the maximum values of NO_X are slightly lower than in the upper figure.

In Fig. 8a complementary structure of the ozone concentration can be found. The comparison demonstrates the principle processes of the nitrogen and ozone budget as they were described in section 3. At this time of the day areas of high concentration of NO_X have low concentrations of ozone. Therefore the regions of high emissions of nitrogen oxides are prominent. In regions of low emissions the concentration of ozone is distinctly higher, due to the photochemical production of the previous day and the transport into these areas. The distribution of the daytime concentration of ozone leeward of the urban area shows a spatial structure which can only be modelled on the basis of regional scale simulations. A more detailed discussion is given in [4].

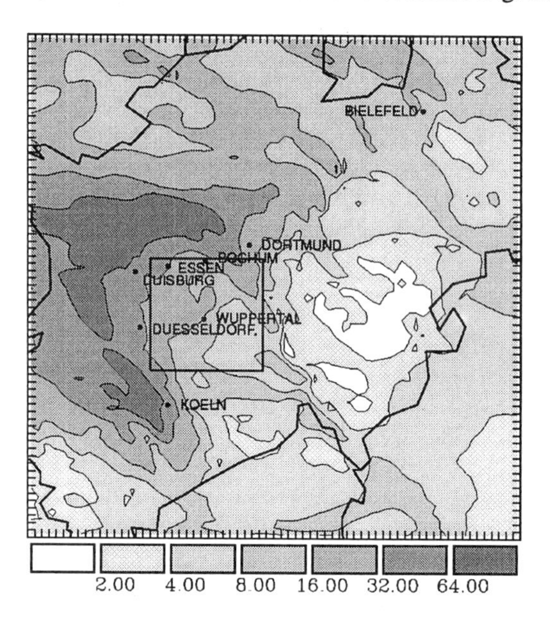

Figure 6: Simulated near surface concentration of NO_X at 6 UTC, 5 June 1996 (in ppbV). Modelling area NRW, 68x68 grid points with approx. 3 km resolution. The inner square depicts the region of the modelling area Ruhr-Wupper.

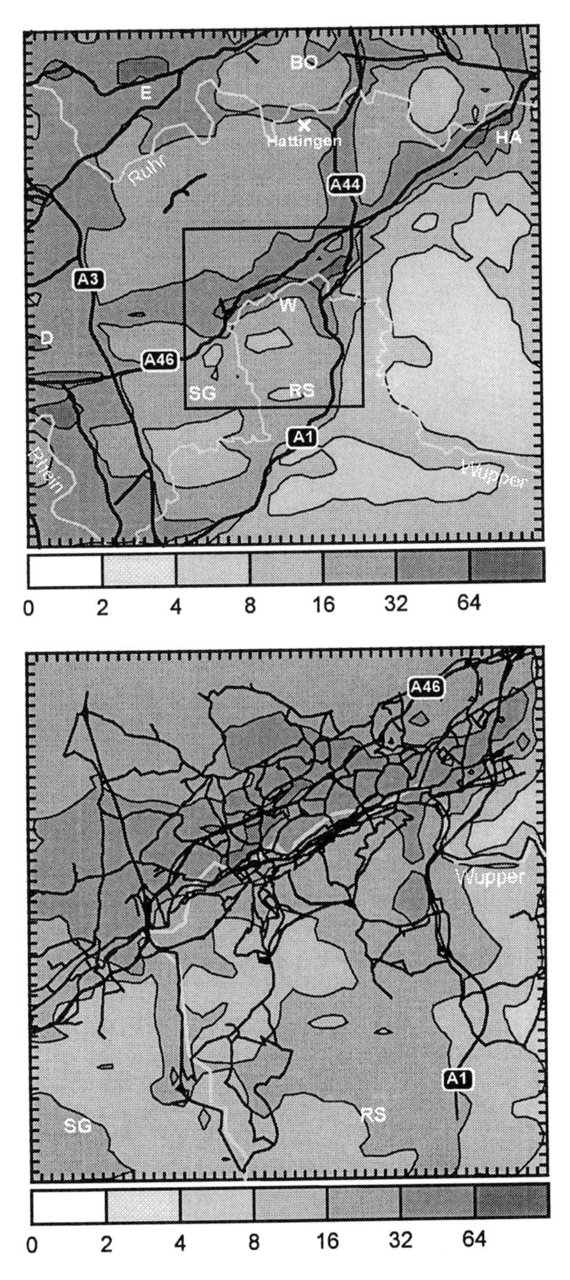

Figure 7: NO_X in ppbV as in Fig. 6, but nested modelling area Ruhr-Wupper, 50x50 km with 1km resolution (top) and modelling area Wuppertal, 17x17 km with 333 m resolution (bottom).

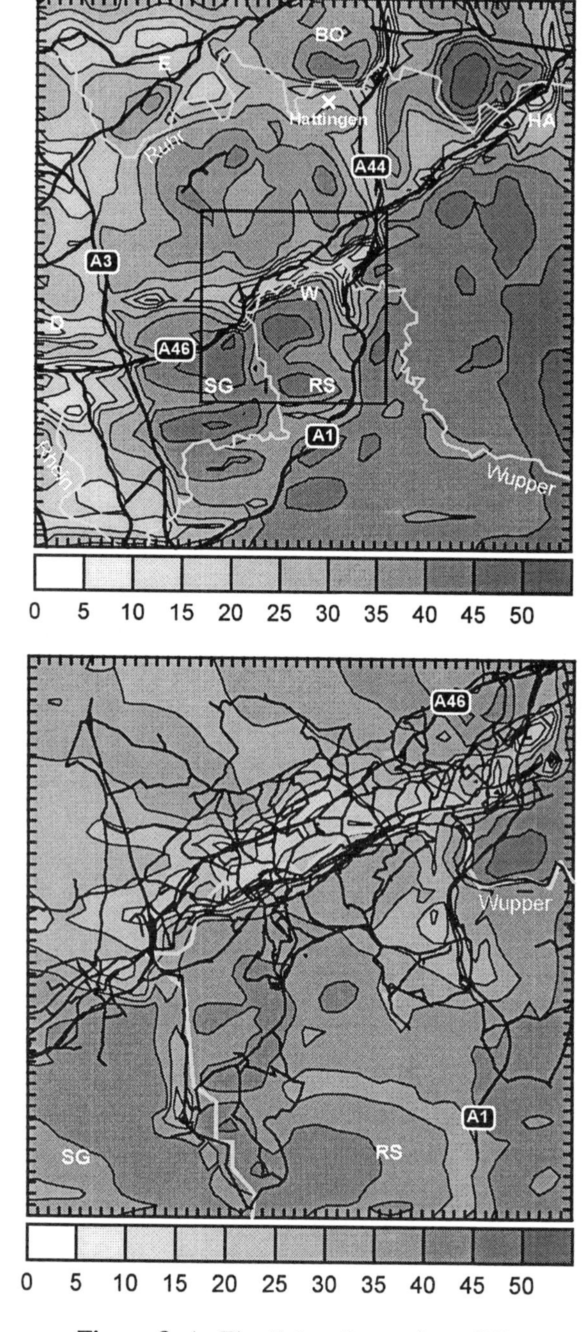

Figure 8: As Fig. 7, but Ozone (in ppbV).

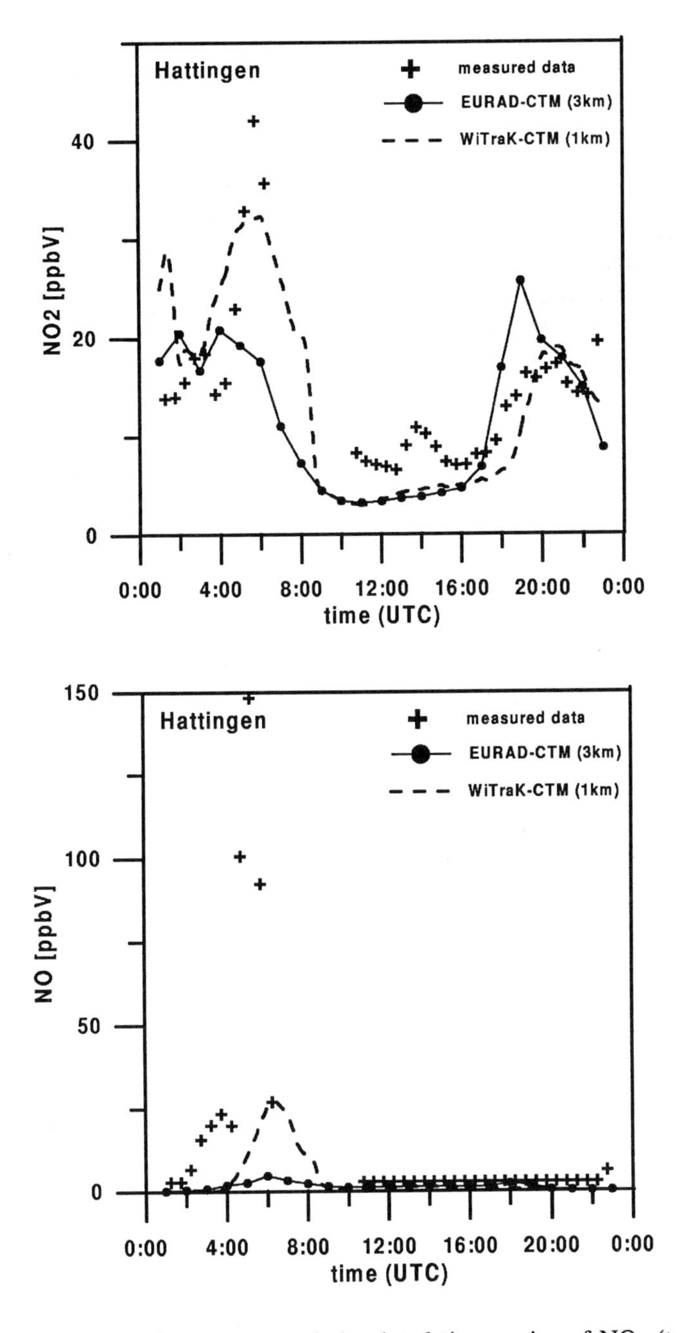

Figure 9: Comparison of measured and simulated time series of NO_2 (top) and NO (bottom) at the LUA-station Hattingen on 5 June 1996.

The amount of the modelled concentrations depends strongly on the spatial structure of emissions and hence on the resolution of the model. The models of the CARLOS system use an eulerian grid in which the concentration is calculated as the mean value of a grid box. Emissions of point or line sources are instantaneously distributed into the volume of the grid resulting in an underestimation of the near source concentrations.

Those effects can be found in Fig. 9 and Fig. 10 at the LUA-station Hattingen (for position see Fig. 7 and Fig. 8). The effects of the emission at the rush hours on the concentrations are better reproduced when modelled with a higher resolution (Fig. 9). For NO, which is directly dependent on the emissions near the source, a resolution of 1 km is too coarse for an accurate representation of the local concentration peak in the morning. During the day NO_X is diluted mainly because of enhanced vertical turbulent mixing. During the evening with reduced turbulence the nitrogen balance is dominated by NO_2. This results from the sufficient concentration of ozone and from the lack of solar radiation.

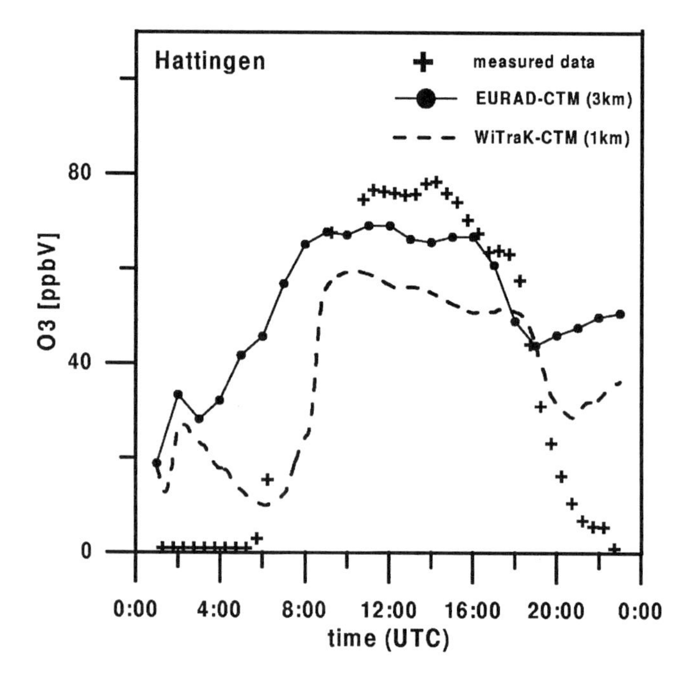

Figure 10: Measured and simulated time series of O_3 at the LUA-station Hattingen on 5 June 1996.

The modelled ozone peaks (see Fig. 10) based on the emission data are too low. The representation of the minima - analogous to the representation of the NO_X-peaks – at sites which are located close to the source is not sufficient due to the resolution of 1 km. An enhanced resolution tends to give better results.

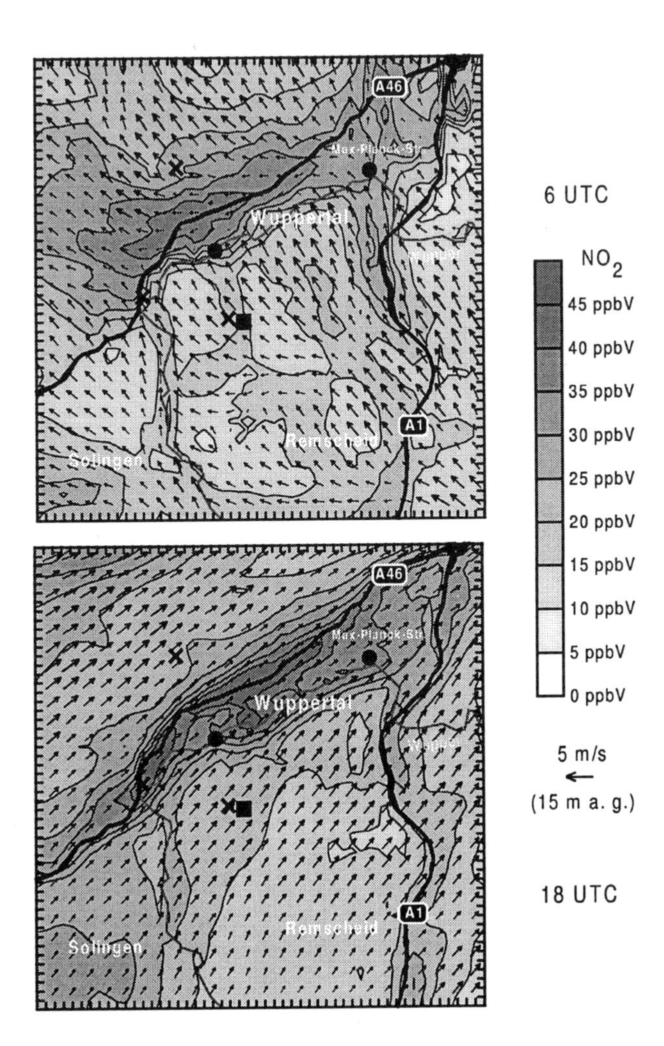

Figure 11: Near surface concentrations of NO_2 and wind field for the inner model area of 17x17 km. Measuring sites are marked by dots (meteorology, NO, NO_2), crosses (meteorology) and a square (ozone).

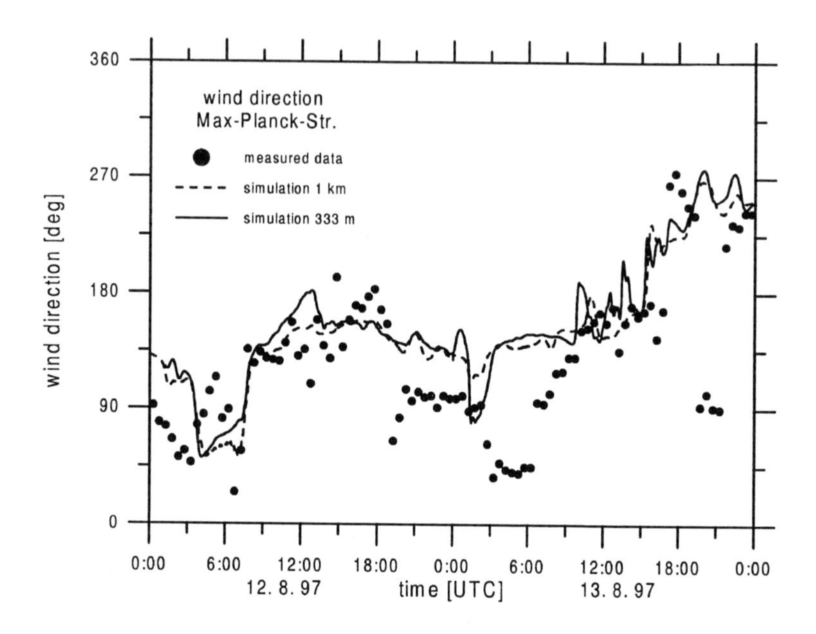

Figure 12: Comparison of measured and modelled wind direction at Max-Planck-Straße.

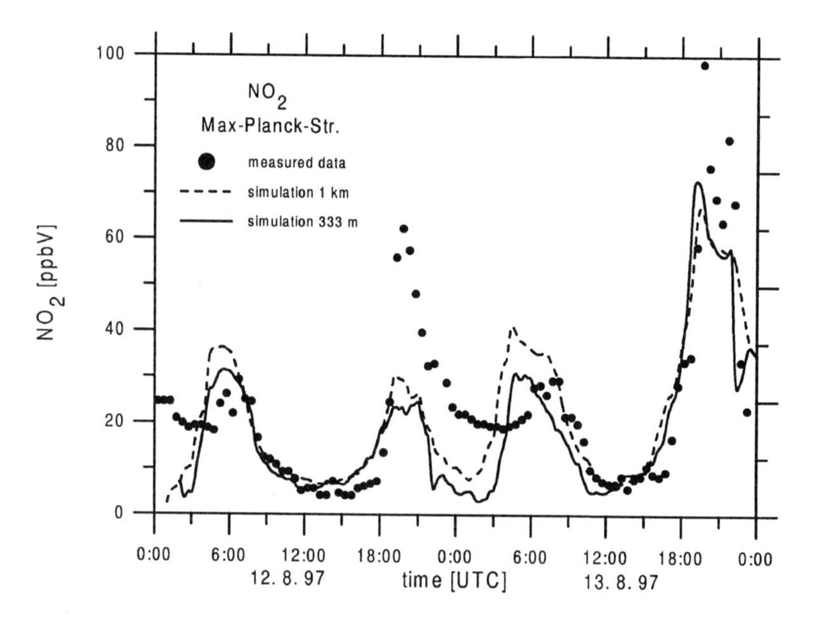

Figure 13: Comparison of measured and modelled concentration of NO_2 at Max-Planck-Straße.

For 1996 no measurements were available for the inner modelling domain. For this reason, the calculations were repeated for a summer smog period in August 1997 on the basis of emission data shown in Fig. 5. Fig. 11 shows the resultant NO_2 concentration and the wind field at 6 UTC and 18 UTC on 13 August 1997. During that day a change of the wind direction from SE to SW occurred. In the morning hours the principle mechanisms are the same as in Fig. 7. After the change to westerly winds polluted air from the Düsseldorf area was transported to Wuppertal. In addition, the flow was nearly parallel to the orientation of the dominant traffic emissions inside the valley. As a consequence relatively high concentrations occur in the evening. Measurements at the site Max-Planck-Straße (for position see Fig. 11) compared with simulated time series are shown in Fig. 12 and Fig. 13.

A close look at Fig. 12 shows that the overall wind direction is modelled quite satisfactorily, but some details, especially during the night, differ between reality and simulation. Although the model tends to reproduce the channelling and katabatic flow inside the Wupper valley during the night this effect is underestimated, especially between 12th and 13th August. This may be an effect of the very structured terrain which is still poorly reproduced by the 333m resolution. Differences in wind directions of 90° of course lead to pronounced differences between predicted and measured concentrations. This can be seen in Fig. 13. Large deviations from the NO_2 measurements occur during the night from 12 to 13 August. During the morning hours the NO_2 values are overpredicted which is due to the location of the measuring site on the top of a 4 floor building leading to relatively low values compared to in-street measurements. Apart from these the simulated concentrations are in good agreement with measurements. The emission data based on dynamic traffic simulations result in similar concentrations as the LUA-data.

The simulations show the principal applicability of the complete hierarchy of models beginning with the simulation of traffic and ending with the prediction of local concentrations. However a detailed look into the extremely difficult mountainous urban Wuppertal area shows that even a resolution of 333m is not sufficient to model all details of the local flow field. But bearing in mind that no measurements were used to constrain the model chain, except for initial meteorological analysis data on the European scale, the results are very encouraging.

5 Outlook

In future projects the CARLOS system will be extended for the purpose of calculating complete climatologies of pollutant concentrations based on emission inventories. In an interdisciplinary approach it will act as a tool which helps to clarify the dependencies between climate trends, economic implications, traffic, emission, local pollution levels, health effects and other related problems.

References

1. H. Hass, *Description of the EURAD Chemistry-Transport-Model Version 2 (CTM2)*, Mitteilungen aus dem Institut für Geophysik und Meteorologie, Universiät zu Köln, **83**., (1991).

2. W. Brücher, *Numerische Studien zum Mehrfachnesting mit einem nicht-hydrostatischen Modell,* Mitteilungen aus dem Institut für Geophysik und Meteorologie, Universität zu Köln, **119,** (1997).

3. W. Brücher, M. J. Kerschgens, R. Martens, H.Thielen and K. Maßmeyer, *Tracer experiments in the Freiburg-Schauinsland area – Comparison with flow and dispersion models*, Meteorol. Zeitschrift, N.F. **7**, 36-40, (1998).

4. C. Kessler, W. Brücher, M. J.Kerschgens and A. Ebel, *Interaction of Traffic-Dependent and other Anthropogeneous Emissions in Case of Pollution Spreading in Strained Areas and Their Surroundings – A Mesoscale Model Investigation*, this issue.

5. W.R. Stockwell, P.Middleton and J.Chang, *The second generation regional acid deposition model chemical mechanism for regional air quality modelling*, J. Geophys. Research, **95**, 16343-16367, (1990).

6. H. Geiß and A. Volz-Thomas, *Berichte des Forschungszentrums Jülich: Lokale und regionale Ozonproduktion: Chemie und Transport*, Berichte des Forschungszentrums Jülich**, 2764**, (1992).

Interaction of Traffic and Other Anthropogenic Emissions in Polluted Regions and Their Environment

C. Kessler, W. Brücher, M. Memmesheimer, M. J. Kerschgens, and A. Ebel

Institut für Geophysik und Meteorologie, Universität zu Köln, Aachener Str. 201-209, 50931 Köln, Germany

Traffic and air quality as aims of the FVU are interconnected and interdisciplinary. CARLOS is a combination of two comprehensive air quality simulation models to calculate chemistry and transport in regional and local scales. Both models apply nesting techniques to describe the influx into the modeling domain. High resolution emission inventories are available for the innermost nest in the large scale, which allow the separate description of traffic and point sources. Results of a simulation for Nordrhein-Westfalen in August 1997 are presented with two nesting levels, increasing the resolution from 27 km in Central Europe to 3 km in the domain of nest2. Comparison of observations at three TEMES-stations with modeled concentrations show a close agreement of ozone and a reasonable reproduction of NO2. Three zerocases are discussed, where industry and point sources as well as traffic are considered in their contribution to ozone production. All cases show a considerable reduction of maximum ozone levels and reductions in mean levels of nitric oxides.

1 Introduction

The interdisciplinarity among topics as diverse as transport of air pollutants and the interests of city planning should not only be connected at the level of expertise but also should be reflected in simulation models.

One of the central topics of the FVU is to assess the effects of traffic on the environment in Nordrhein-Westfalen (NRW) and to create a network of interdisciplinary research institutes participating in computer aided modelling of traffic. One of the projects is aimed especially towards better understanding of air quality in relation to road traffic emissions. Together with all partners reduction scenarios will be evaluated, taking into account effects such as traffic redirection and changes in mobility.

Emissions of traffic, industry, power plants and other sources lead to changes in the combination of species in the air, commonly called air pollution. In anticyclonic situations emitted trace gases can accumulate in the planetary boundary layer and with the attendant high insolation ozone will increase to critical levels. Ozone is a product of precursor species as NOx (nitrogen oxides) and VOC (volatile organic compounds) and radiation with short wavelength. The most recent example of a photosmog episode in Germany has occured in the second week of August 1998. Traffic restrictions have been mandated but were only partially followed. The discussion on ozone reductions by emission changes will be fueled again by the phasing out of the german "Ozonverordnung", end of 1999. The "Heilbronn/Neckarsulm–Ozonversuch" [1] tried to establish effects of traffic reductions in a 20 x 20 km^2 area. The resulting changes in ozone concentrations could not be clearly attributed to the measures (lower speed on interstates, restrictions on city traffic) since the area considered and the amount of precursor emissions affected by the restriction have been too small, compared to the scale of ozone buildup. However, the expected extended traffic jams and reduced mobility did not take place and the experiment proved the feasibility of such restrictions. Unfortunately we might not see another "real" experiment like this one with an enlarged area of restrictions. Simulations, therefore, have to assist in determining the effects and in quest of the optimum or best possible realization of emission scenarios. Photochemical models provide a scientifically well-founded basis for the assessment of emission scenarios. Three-dimensional models are the least limited models, particularly for control strategy analysis of multi-day episodes [2].

A further topic of the FVU is to build and apply tools to study experimentally the widespread effects of traffic. Computer simulations offer a large field of possibilities, which is yet restricted by the widely distributed research facilities and the many facets of traffic. Official sites in Nordrhein-Westfalen have already collected many datasets necessary for the various tasks of modelling, thus adding to the already strong motivation of small scale studies.

The FVU is a step towards the "virtual institute" combining the expertise of traffic planning as well as atmospheric science and business administration. A collection of tools, namely CARLOS, is a product of the FVU and has been applied in this paper. Section 2 gives a short description of the large scale part of CARLOS, the small scale part being described in [3]. Compilation of data is given in section 3, and in section 4 results are presented for several resolutions.

2 Model Description

CARLOS (Chemistry and Atmospheric Transport in the Regional and Local scale) is a suite of nested models consisting of EURAD (European Air Pollution Dispersion Model) and WiTraK. Both are complex Eulerian models with high vertical resolution and one-way multiple nesting. EURAD itself denotes three parts:

- Meteorological driver MM5 [4],

- Emission preprocessor EEM [5],

- CTM2 [6,7], the chemistry-transport model.

Both WiTraK and EURAD calculate chemical transformation with the kinetic reaction model RADM2 [8] (21 anorganic and 42 organic species, 157 reactions). In CTM2 gasphase-chemistry is calculated with an implicit sparse-matrix solver, ROS2 [9], which ensures full conservation of nitrogen. CARLOS uses a detailed description of the vertical dimension with 15 layers. The first layer has a thickness of app. 40 m and the upper layer ends at 100 hPa (ca. 16 km).

With the combination of two comprehensive models, CARLOS is especially suited for calculations which span large scale differences. In this study the horizontal resolution was increased fourfold from 27 km to 333 m with a model interface from 3 km to 1 km. Due to the independent development of CARLOS both parts are adapted to their respective scale, describing pertinent processes with suitable algorithms. The FVU is focussed on traffic at local to regional scales, i.e., the main modeling area is the Bundesland Nordrhein-Westfalen.

To describe air masses flowing into this densely populated region a network of wind and pollutant concentrations is necessary to get information on spatial and temporal structure. Observations at air quality stations are limited to the near surface and only few constituents are measured. Therefore, we have to use large scale models to describe additionally the fate of hydrocarbons and photo-oxidants and some slow radicals for the full model height. Of course, these problems are also present with large scale computations, however, the influx into the modeling domain, e.g. at sea, can be described adequately with background values. This is valid, if the influence of far away boundaries is small compared to the contribution of emissions inside the domain and during meteorological conditions governed by high-pressure systems typical for photosmog episodes. Methods to create a suitable initial state in CTMs, based on observations, are being developed, eg. variational data assimilation [10].

Another issue in atmospheric modelling concerns the resolution of the model grid, expressed as length of a square cell. In order to adequately describe the impact of traffic on the local environment, i.e. to resolve interstates, it is necessary to use grid lengths as small as a few hundred metres. Combined with the need to move boundaries far away to include the history of air masses (length of domain at 1000 km) this would lead to yet unmanageable model sizes of billions of grid cells and computing times of several months for a single day.

Nested models offer a solution to the difficulty of size vs. resolution. The study area is embedded into larger domains, these into still larger ones and the grid length is increased, e.g. by a factor three. Calculating top down provides each smaller area with proper boundary conditions without the need of, at present, unavailable computer ressources.

3 Input Data

The simulated episode started 7 August and ended 14 August 1997. Data for three model domains are specified as inputs to the meteorological model and the chemical transport models. The term input data is not precise, as results from the coarse model act as input to the first nest and so on. The study domain is centered on NRW in the nest2, surrounded by nest1 with parts of Germany, the Netherlands and Belgium. The coarse domain encloses most of Central Europe. Tab. 1 lists the model dimensions and computing times for a two day episode.

Meteorology is computed starting with initial and boundary conditions for the coarse domain on the basis of ECMWF (European center for medium range weather forecast; Reading, UK), subsequent nests are computed within MM5. The output of MM5 is preprocessed to CTM2-format and deposition velocities and rates for clear sky photolysis are added. Photolysis correction factors are computed inside CTM2 to account for clouds and atmospheric transmission losses. An example of the computed wind is given in Figs. 1 and 2. A stationary anticyclone above northern Germany slowly transported moderately polluted air from south-east Germany (Fig 1). During the day (Fig. 2) low wind speed prevailed and wind direction in NRW was south. The end of the ozone episode is reached with the increase and change of wind direction in the night.

Emission datasets were prepared with two different approaches: In the coarse and nest1 domain EMEP data were processed with the EEM: yearly emission totals were regridded from the 150x150 km^2 EMEP-grid, provided with time functions for month, type of day and diurnal variation and distributed according to landuse and population density. In the nest2-domain high resolution datasets with a 1x1 km^2 grid size were available from the regional environmental agency (Landesumweltamt Essen). The traffic inventory also has a time resolution of one hour for a workday, saturday and sunday. Point sources and small industry emissions are given as monthly means. The latter datasets were regridded from Gauss-Krueger-coordinates to the Lambert conformal system, which is the coordinate system of CTM2. For grid cells, which are outside of NRW, information from the nest1-inventory was added. In all model inventories biogenic emissions were calculated using the temperature simulated by MM5. Tab. 2 gives the sum of daily emissions into nest2 and Fig. 3, left, shows an overview of the spatial distribution of the daily sum of emissions of NOx for 9 August 1997. The distribution of VOC (Fig. 3, right) shows the contribution of forested areas and high emission fluxes inside cities caused by low driving speeds and congestions with typical low engine load. On the other hand, high engine load and temperature lead to the typical NOx-pattern of interstates in NRW. No additional data sets were provided for the CTM2, i.e. modelled concentrations are not nudged or otherwise adapted to measured concentrations. The latter were only applied to post-simulation model performance tests.

4 Model Experiments

The effect of an emission source or source category on calculated concentrations can be computed by using a zerocase. In the dataset for the zerocase the emission flux due to the specific source category is subtracted from the basecase dataset, i.e. the source is shut off. The differences in concentration levels then result solely from the excluded emission source. However, results of such a zerocase should not be taken as experiment in reality e.g. simulation of no traffic in one region cannot be calculated by simply excluding traffic from the dataset. Rerouting of traffic and increase in congestions outside the region would have to be considered as well. Zerocases denote an upper limit of changes where cause and effect are combined linearly. Furthermore, they allow the detection of response times of the system considered. Three model experiments have been considered: Excluding production and power generation, excluding traffic and exluding all anthropogenic emissions. In all cases meteorological parameters and biogenic emissions remained unchanged to the basecase.

5 Results

Basecase Simulation

The basecase, excluding two days for spin-up, was simulated from 9 August 1997 to 14 August 1997. At 6 UTC the low mixing layer height prevents efficient vertical mixing of emissions and leads to a build-up of NOx-concentrations (Fig. 4, left). From Munich to Rotterdam, elevated NOx-levels are visible in a half crescent shape, where ozone concentratons are less than 30 ppbV (Fig. 4, right, coarse and in parts also visible in nest1) due to the titration by NO. In nest2 the industrialized center north-west of Cologne and power plants between Cologne and Aachen form a distinct NOx-plume. Ozone levels above 30 ppb indicate the residual of the night in clean and elevated rural areas as North Sea, Eifel and Rothaargebirge.

After the increase of the morning mixing layer height, at 14 UTC (Fig. 5, left), NOx levels have markedly decreased with spots around the large cities and between Cologne and Birmingham. NOx in nest2 shows lower levels as in the coarse domain and is oriented with the southern wind. A band of elevated ozone concentrations reaches from Milano to Schotland, partly along the Rhine valley, with local peaks e.g. at Basel and in NRW (Fig. 5, right). Ozone levels are higher in nest1 with local minima close to cities. A northern orientation is also visible in the distribution of ozone in nest2 where strong point sources show titration of ozone. The effect of these point sources is lost in the lower resolution of nest1 and coarse. A region with high ozone concentrations in the south east of NRW coincides with high biogenic emissions (Fig. 3, right) from forests in Hessen.

Model performance can be compared to reality, e.g. observations at air quality stations. Ideally, observations should be spatially interpolated to the model grid to be fully comparable. Fig. 6 gives an example of a summer ozone observation at Duisburg-Walsum for the period of three months. The length of the roses indicate the mean concentration at the particular wind direction. Mean wind speed is added in the plot to show the strength of transport at the direction. Ozone concentrations are high (Fig. 6, left) with south-easterly wind, typical for anticyclonic situations in that region. With low wind speed and weak mixing photochemically produced ozon can accumulate. The strong orientation of the NO2 rose (Fig. 6, right) in northern direction denotes the local transport direction of emitted NOx from south.

In Fig. 7 simulated concentrations in nest2 of ozone and NO2 are compared to results from the TEMES-station Duisburg-Walsum. The ozone concentrations during most of the day, as well as the maximum values, are reproduced quite satisfactorily. The deviations at night indicate local influences other than those, which can be reproduced fully in the model, e.g. the amount and diurnal variation of NO2 measurements represents the near influence of traffic. Stations at Bielefeld and Köln (Figs. 8 and 9) provide a similar comparison with a good reproduction of ozone and differences in NO2. Note that around Bielefeld (BIEL in Fig. 10) the spatial emission pattern is different compared to that of Köln or Duisburg and small scale models (WiTRaK) will improve the description to an even closer agreement.

Zerocase: without Industry and Powerplants

In the current inventory for NRW the contribution of industry and powerplants ('w.o. industry' in the remainder of this text) to the emitted flux is modelled as a large number of point sources. Results of a simulation with the reduced inventory are given in Fig. 10 (upper row) for 13 August 1997. Computed concentrations for the basecase are subtracted from the results of the zerocase. The differences in NOx to the basecase are finely structured due to the distribution of the point sources and, caused by the midday southern wind, oriented south-north. Reducing NOx emissions can also slightly increase NOx-concentrations by reduced HNO3-formation (see also Fig. 11, lower right). In some grid cells in the vicinity of NOx-rich sources, removal of NOx can lead to less titration of ozone, i.e. net production occurs, further downwind this process is reversed when air masses are chemically "aged".

Zerocase: without Traffic

The dense network of roads in NRW leads to considerable changes in both NOx and ozone (Fig. 10, middle row). In contrast to 'w.o. industry' all traffic emissions are emitted in the lowest level of the model. At sites, where also strong point sources emit, the changes in NO2 are less marked. Once again the transport direction is south-north and along the southern boundary of NRW influx of ozone dominates, i.e. changes are close to zero. Also, all changes are negative in the domain: the composition of traffic emissions is more VOC-rich than e.g. those of power plants with high efficiency. Thus, the ozone reducing effect of NOx is counterbalanced by reducing the ozone-precursor VOC.

Zerocase: without Industry and Traffic

This combination of the aforementioned zerocases repesents the situation, where no anthropogenic sources are considered in the inventory (Fig. 10, bottom row). Reductions in both ozone and NOx are generally more distinct than in considering only one source category. Increase in ozone concentration compared to the case 'w.o. industry' extends to smaller areas as mixing with less polluted air lessens the effect of strong NOx-sources.

Zerocase: Summary

In Fig. 11 mean diurnal variations are shown for the near surface of nest2 for 13 and 14 August 1997. A distinct difference in all species between the first and second day is caused by higher wind speed and vertical mixing on the 14 August 1997. Low wind speeds are often correlated with high pressure and weak vertical exchange, where pollutants can accumulate in the mixing layer. This leads to a sharp NO2- and NO-peak on the 13. HNO3 is a precursor for acid deposition and reacts differently from the other species shown. HNO3 is produced by reaction of NO2 and OH, the latter being a prime oxidant and 'cleansing agent' of the atmosphere. With better mixing of upper air at night low OH-nighttime reservoirs might be replenished from the free troposphere above the nighttime inversion and can thus react with NO2.

As already discussed exclusion of industry induces only moderate changes in ozone at 14 UTC. The diurnal variation in Fig. 11 (upper left) indeed shows at night an increase in mean ozone concentrations and only negligible changes during the peak. The other cases are more marked in each direction. The consistent nocturnal increase of ozone in all cases occurs because at night reaction with NO is the sole important sink of ozone, i.e. removal of NOx in polluted areas must lead to an increase. At daytime, however, also the amount and composition of hydrocarbons affects the net production rates. All zerocases also affect the amplitude of mean diurnal ozone concentration, i.e. emissions increase the span between day and night.

This finding is more pronounced if maximum concentrations are considered. In Fig. 12 ozone concentrations between 0 UTC 13 August and 23 UTC 13 August were scanned for absolute maximum values in each case for all gridcells. The horizontal coordinate of one single dot is the maximum value in the basecase and the vertical coordinate is the change in maximum ozone concentrations at the same grid cell in the zerocase. A moving average with a mean of 100 values is added as line. In all cases emissions lead to a strong increase in maximum ozone values. Below a threshold of ca. 60 ppb ozone concentrations are not affected in this domain.

Mean concentrations of NO2 and NO, compared to the basecase, decrease considerably in all cases and at all times and once again show, that the local behaviour of nitric oxides is different than that of ozone, which is produced "elsewhere". Maximum concentrations of ozone, however, are clearly the result of local emissions and are thus affected by regulations.

Acknowledgements

This work was funded by the Ministerium für Wissenschaft und Forschung (MWF) of Nordrhein-Westfalen. We also gratefully acknowledge permission to use extensive datasets from the Landesumweltamt NRW including the TEMES-network. Last not least the support of the Bundesministerium für Bildung und Forschung is essential to continue and further expand our work with EURAD. The preparation of meteorological input data for MM5 has been supported by Prof. P. Speth, Ilona Stiefelhagen and Birdie Roeben, Universität zu Köln. MM5-Simulations have been carried out by H.J. Jakobs, Universität zu Köln. Our results would still be hidden in unimaginative input datasets but for the extensive computing support in the Forschungszentrum Jülich, NRW, in particular ICG2, ICG3 and ZAM and also the RRZK in Köln providing a blindingly fast network.

References

1. Umweltministerium Baden-Württemberg,.Modellversuch Heilbronn/Neckarsulm. Erste Ergebnisse und Erfahrungen, (Stuttgart, 1995).

2. J. Kuebler, J.M. Giovannoni, A.G. Russell Atm. Env. 6, 951-966, Eulerian modelling of photochemical pollutants over the Swiss plateau and control strategy analysis, (1996).

3. W. Brücher, Ch. Kessler, M.J. Kerschgens and A. Ebel, Simulation of traffic induced air pollution on regional and local scales, this volume.

4. G. Grell, J. Dudhia and D. Staufer, A description of the fifth-generation Penn State/NCAR mesoscale model (MM5), Technical Note NCAR/TN-398+STR, (National Center of Atmospheric Research, Boulder, Colorado, 1994).

5. M. Memmesheimer, J. Tippke, A. Ebel, H. Hass, H. Jakobs and M. Laube, On the use of EMEP-emission inventories for European scale air pollution modelling with the EURAD model, in Proc. of the EMEP-workshop on Photoxidant modeling for long-range transport in relation to abatement strategies, Berlin 16.-19. April 1991, (Berlin, 1991).

6. H. Hass, Mitteilungen aus dem Institut für Geophysik und Meteorologie, Universität zu Köln, 83, Description of the EURAD Chemistry-Transport-Model 2 (CTM2), (Köln, 1991).

7. A. Ebel, H. Elbern, H. Hass, H.J.. Jakobs, M. Memmesheimer, M. Laube, A. Oberreuter, G. Piekorz, Simulation of chemical transformation and transport of air pollutants with the model system EURAD, in: Transport and Chemical Transformation in the Troposphere, eds. P. Borrell et al., Vol. 7, Tropospheric Modelling and Emission Estimation, Springer Verlag, (Berlin , 1997).

8. W.R. Stockwell, P. Middleton, J. Chang, X. Tang, J. Geophys Research 95 D10, 16343-13367, The second generation regional acid deposition model chemical mechanism for regional air quality modeling, (1990).

9. J. Verwer, E.J. Spee, J.G. Blom and W. Hundsdorfer, CWI-Report MAS-R9717, A second order Rosenbrock method applied to photochemical dispersion problems, to appear in SIAM J. Sci. Comput., (1997).

10. H. Elbern, H. Schmidt and A. Ebel, J. Geophys. R., 102, D13, 15,967-15,985, (1997).

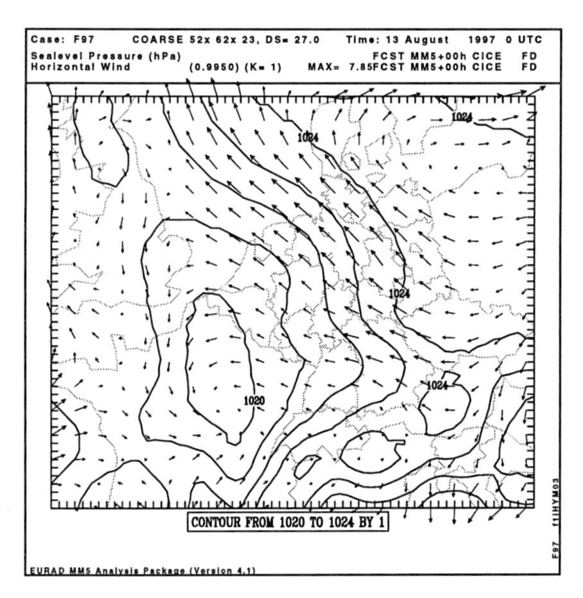

Figure 1: Windfield and sealevel pressure at 0 UTC, 13 August 1997. Coarse domain with Nordrhein-Westfalen in the centre.

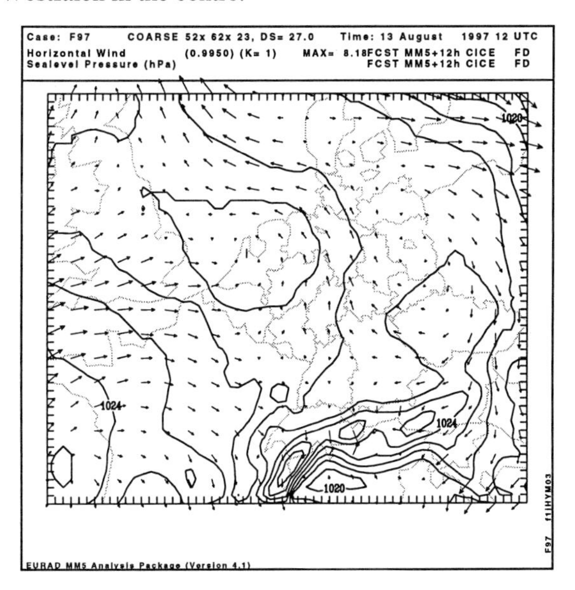

Figure 2: Windfield and sealevel pressure at 12 UTC, 13 August 1997. Coarse domain with Nordrhein-Westfalen in the centre.

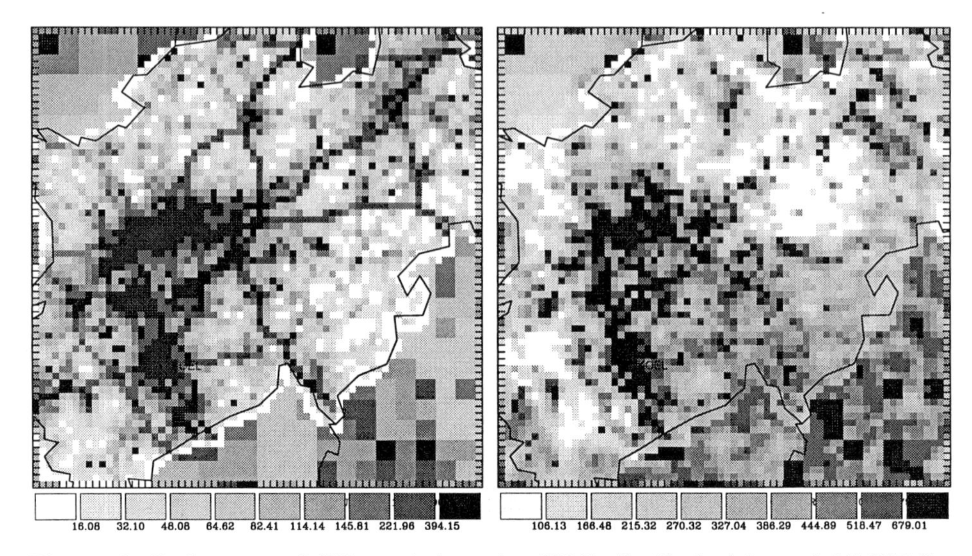

Figure 3: Daily sum of NOx-emissions (as NO2) [kg/day] right and NMVOC-emissions [kg/day] for 13 August 1997. The greyscale is in 10 percentiles: lowest printed legend is 10. percentile, last printed legend is 90. percentiles.

	imax	jmax	kmax	dx, dy	CPU [secs]
Coarse	61	51	15	27 km	7377
nest1	55	55	15	9 km	6807
nest2	70	70	15	3 km	12816

Table 1: Grid dimension and size of the model domain. CPU-time for two days on a vector computer Cray T90, one processor.

Species	Emissions [to/day]
Nox	1391
CO	3937
VOC	1119
ISO	1186

Table 2: Sum of emissions for 13 August 1997 for the nest2 domain in Fig. 3.

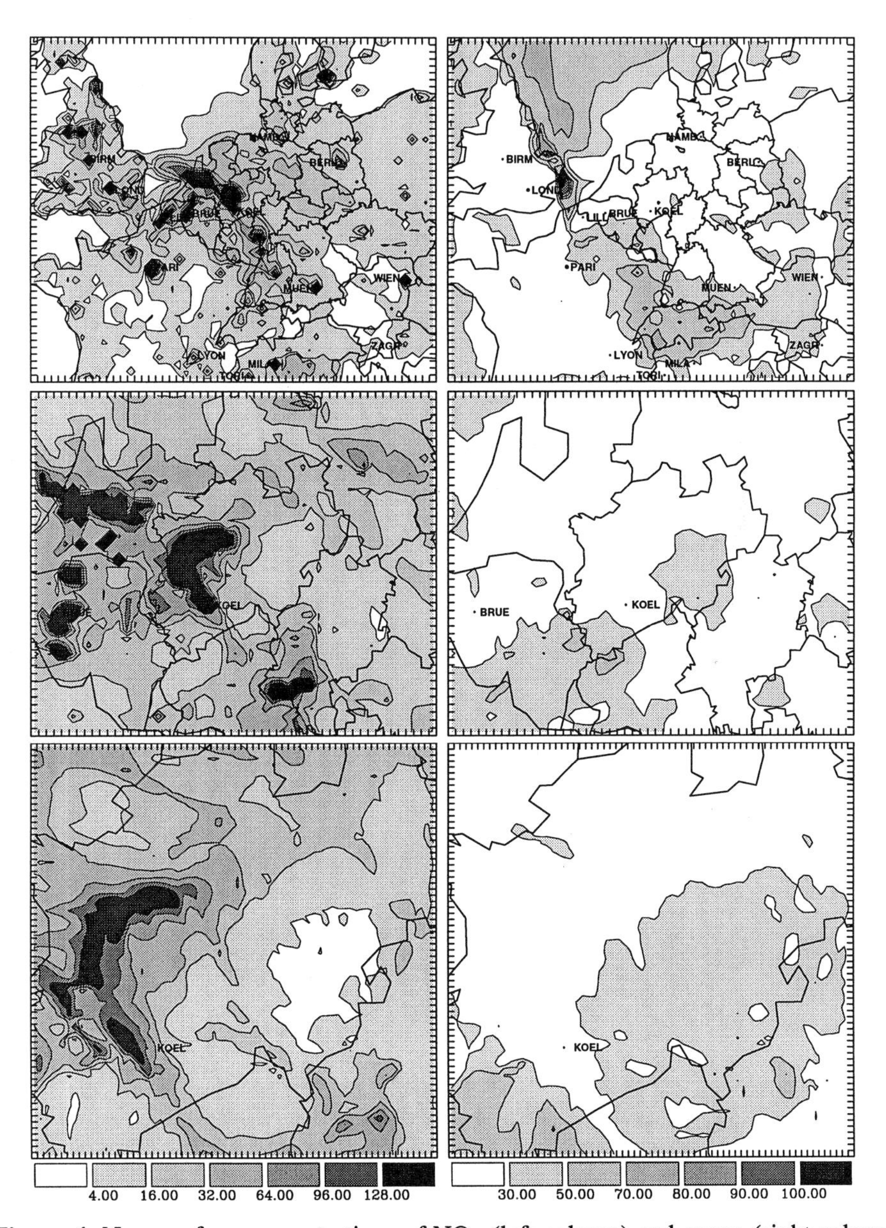

Figure 4: Near surface concentrations of NOx (left column) and ozone (right column) at 6 UTC 13 August 1997, simulated with EURAD-CTM2. From top to bottom: Coarse domain, nest1 and nest2.

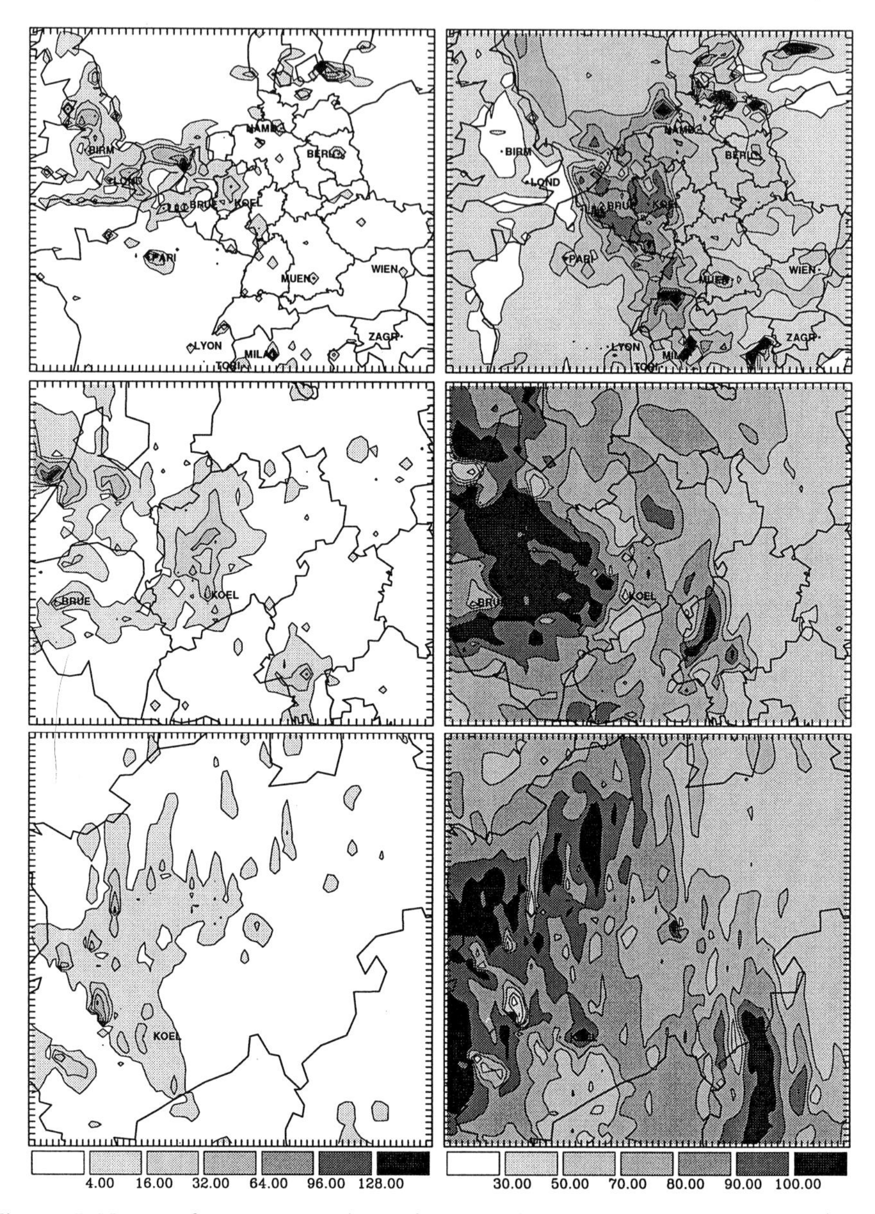

Figure 5: Near surface concentrations of NOx (left column) and ozone (right column) at 14 UTC 13 August 1997, simulated with EURAD-CTM2. From top to bottom: Coarse domain, nest1 and nest2.

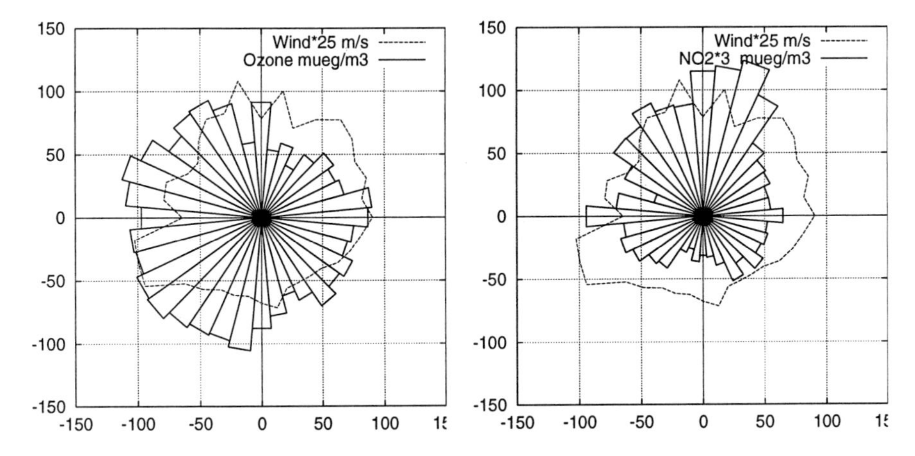

Figure 6: Ozone and NO2 for the hours 12 to 18 MEZ for three months June till August 1997 at the TEMES-station Duisburg-Walsum. Shown are mean values of the concentrations oriented in the winddirection (north at top), i.e. peak in North caused by southern wind. Ozone indicates, that elevated ozone levels are correlated with eastern wind.

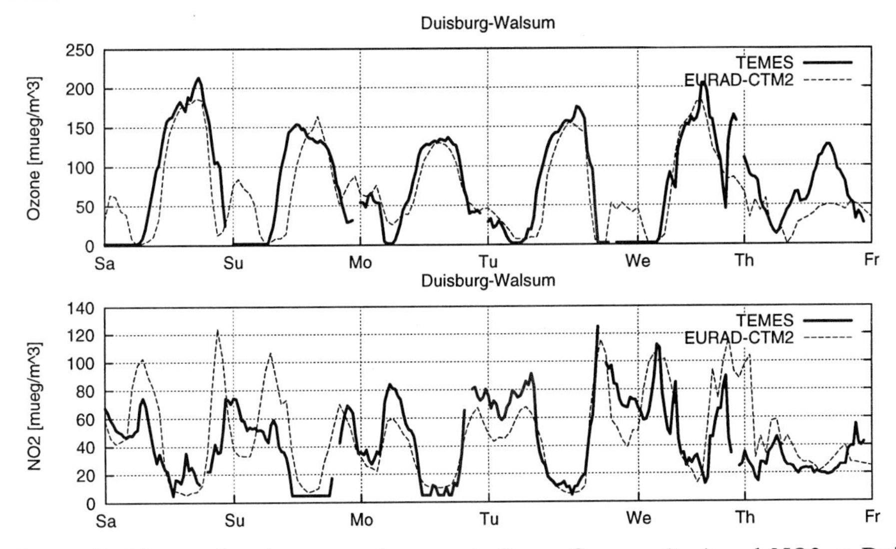

Figure 7: Measured and computed concentrations of ozone (top) and NO2 at Duisburg-Walsum for 9 - 14 August 1997. Measured concentrations at a height of app. 10m, computed concentrations at app. 20 m. The station Duisburg-Walsum is part of the NRW-TEMES network.

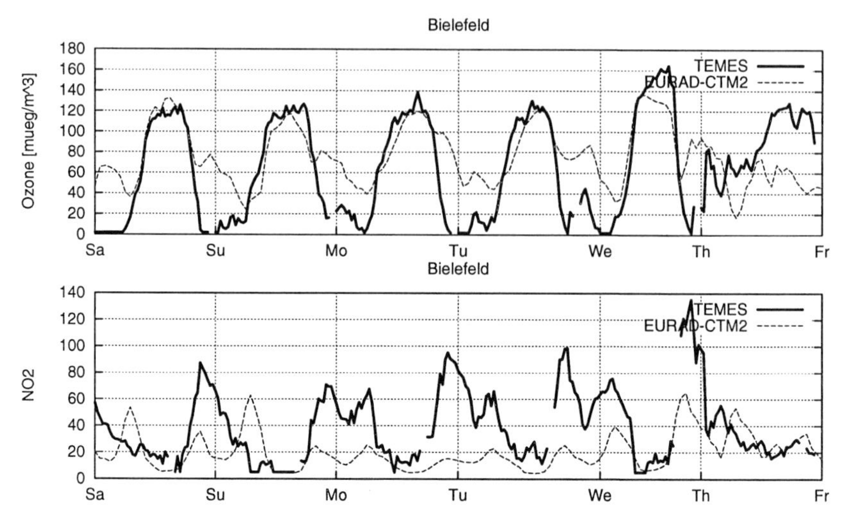

Figure 8: Measured and computed concentrations of ozone (top) and NO2 at Bielefeld for 9 - 14 August 1997. Measured concentrations at a height of app. 10m, computed concentrations at app. 20 m. The station Bielefeld is part of the NRW-TEMES network.

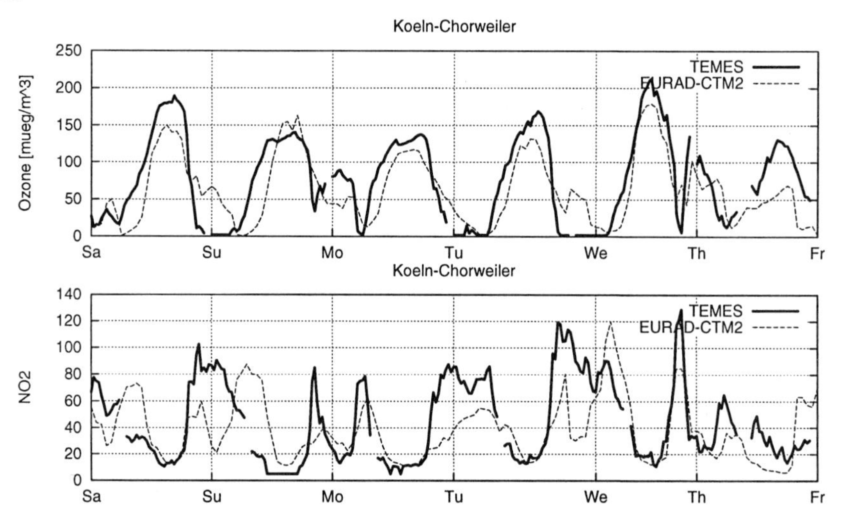

Figure 9: Measured and computed concentrations of ozone (top) and NO2 at Koeln-Chorweiler for 9 - 14 August 1997. Measured concentrations at a height of app. 10m, computed concentrations at app. 20 m. The station Köln-Chorweiler is part of the NRW-TEMES network.

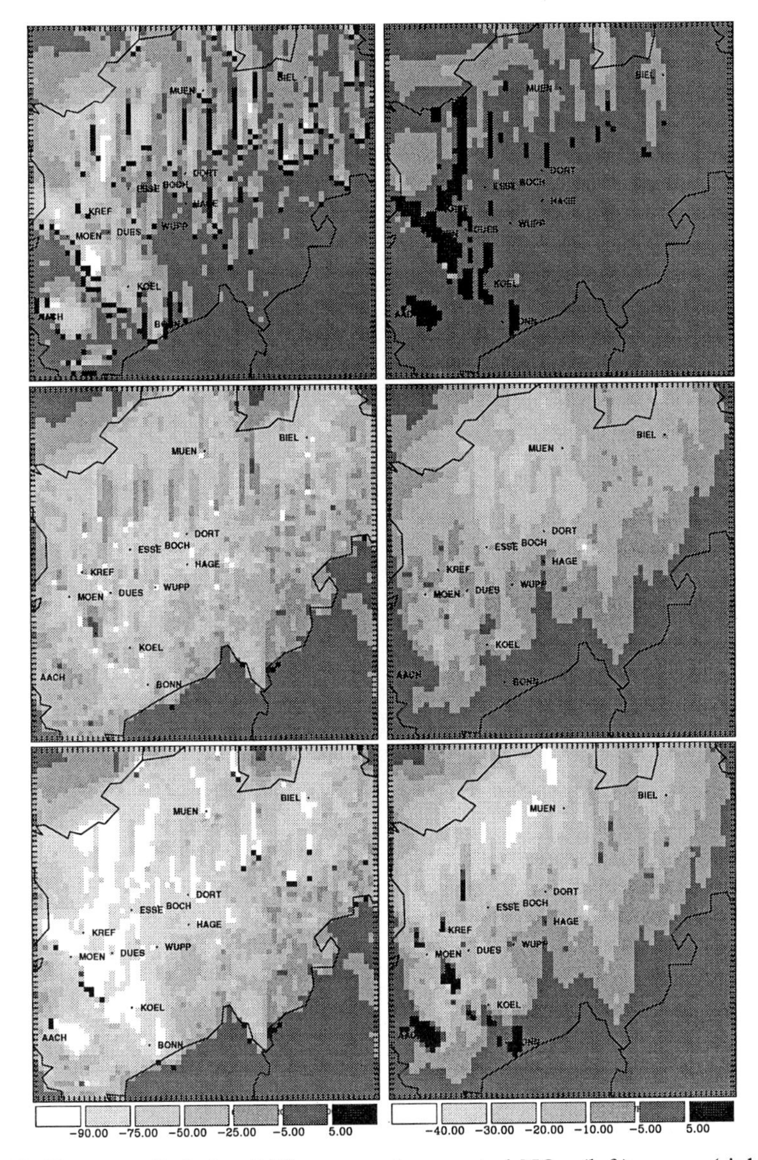

Figure 10: Zerocase: Relative Differences of computed NOx (left) ozone (right) concentrations to basecase [%] at 14 UTC 13 August 1997. Upper row: without industry and power plants, middle row: without traffic, bottom row: without anthropogenic emissions.

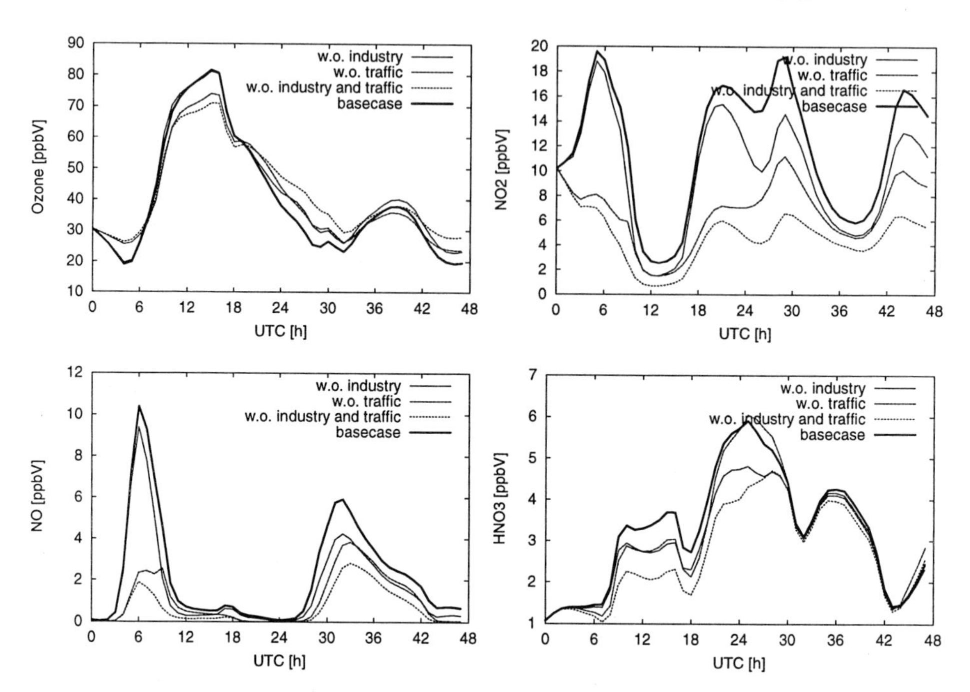

Figure 11: Mean simulated concentrations, nest2, at near surface (app. 20m) for the nest 2 domain (NRW). Ozone and NO2 (upper row), NO and HNO3 (lower row) in [ppbV]. Shown are basecase and three zerocase computations, 13 and 14 August 1997.

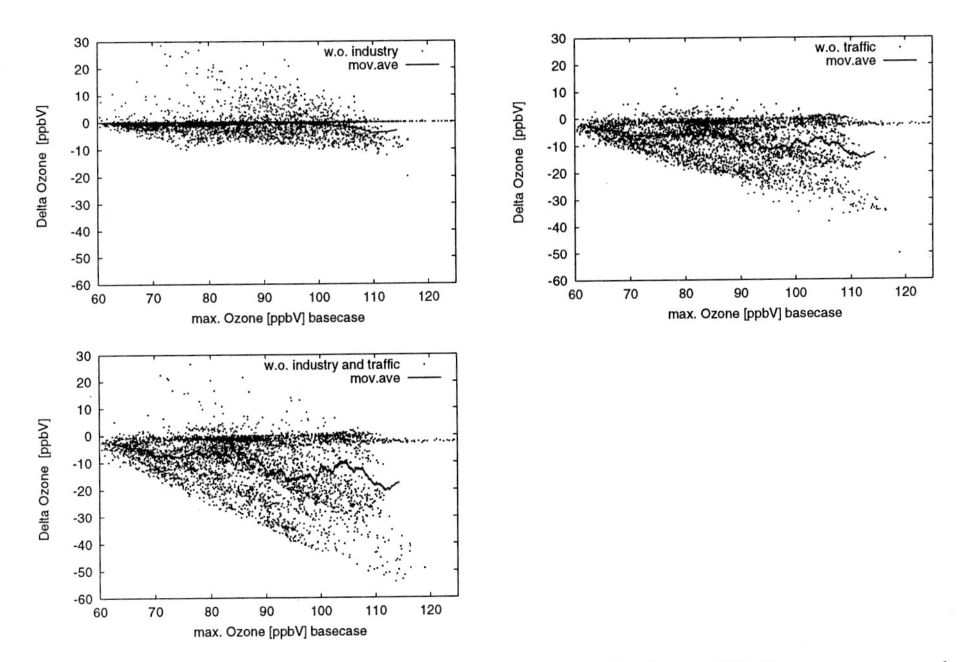

Figure 12: Reductions of max. ozone concentration [Delta ppbV] for zerocases and max. ozone concentration of basecase [ppbV], 13 August 1997 in nest2, near surface. Moving average with 100 points as line.

Time Soluted Assessment of Traffic Impacts in Urban Areas Based on Dynamic Traffic Simulation

F. Huber and S. Kaufmann

Bergische Universität-Gesamthochschule Wuppertal, Fachbereich Bauingenieurwesen, Lehrgebiet Umweltverträgliche Infrastrukturplanung, Stadtbauwesen (LUIS), Paulus-kirchstrasse 7, 42285 Wuppertal, Germany

The task of LUIS within the research cooperative is the assessment of time soluted traffic impacts in towns. For this, time dependent limit values for a compatible traffic volume had to be developed. They are defined as the maximum number of vehicles per street section for the period ½ h. Furthermore, urban space types and their sensitivity had to be determined. By that, in the future it will be possible to integrate aspects of compatibility in the development of control strategies for the motorized individual traffic.

1 Introduction

The Institute of Infrastructure and Urban Planning, University of Wuppertal (LUIS) has the task to assess the traffic volume simulated by the research partners, on basis of the specific sensitivities of spatial structures in towns. By that, in the future it will be possible to integrate aspects of compatibility in the development of control strategies for the motorized individual traffic (e.g. traffic management, modal choice and route control in densely populated areas, telematics applications) beyond the efficiency of a road.

The contribution of LUIS covers the four areas:

- integration of the spatial structures (networks; building- and use structures, sociodemographic basic data),
- sensitivity assessment of the of spatial structures,
- assessment of traffic impacts on humans,
- development of the space-related information system LUISE for the calculation, assessment and presentation of traffic impacts in larger networks of the motorized individual traffic, based on the sensitivity of spatial uses.

With the general condition of high time dependent and spatial solution of the traffic, the assessment of impacts deals with new demands. But it also opens completely new possibilities and chances. These are situated primarily in the conclusion of the dimension "time" into the view.

As a result of the netwide time dependent analysis of the traffic loads and their impacts, a modified planning perception of the pulsating traffic in the city arises. New questions come about, e.g. how short strong peak loads compared with long lasting impacts have to be judged. For their answer and to define a compatible volume of traffic in cities, the development of spatially more differentiated and time resolved thresholds was necessary. They are defined as the maximum number of vehicles/section for the period of ½ h. In the future it shall be possible to pass the information, whether sections in urban nets have exceeded limit values, to dynamic navigation systems, so that the links are not taken up to route recommendations for such a long time, until the overloading is reduced below the threshold values. The driver instead may get information about alternative "compatible" routes or the use of public transport, in order to reach the required destination.

This measure appears necessary, because dynamic navigation systems recommend in the case of congestions in the superordinate net (motorways), usually subordinate net sections (in towns) as alternative routes. It was proven, that by the substantial quantity effects outgoing from these systems, in the future a strong increase of superordinate traffic in the urban road system follows.

The assessment of the traffic impacts concentrates on major roads in built-up areas with connecting function (Guidelines for the design and alignment of streets –part nets, road categories C) [1]. Streets with predominant residence and development function are not taken into consideration, since they should generally be excluded from through traffic.

2 Use and Sensitivity of Urban Spaces

Since the definition of maximum compatible traffic volumes on a section is based on the sensitivity of adjacent uses, first a classification pattern for the close-to-reality figure of street-space constellations had to be developed. The types of buildings and the land use are determined in the urban land-use planning based on specifications in the german "Federal Building Code (FBC)" (Baugesetzbuch) and the "Federal Utilization Ordinance (FUO)" (Baunutzungsverordnung). This technically common use categorization is maintained for compatibility reasons. Since this systematization is not based however constantly on the sensitivity of the urban spaces, new or extended definitions had to be developed for the following area types.

Mixed Use Areas

The classification of spatial uses in the context of urban land-use planning is determined in the FUO by the given quality of use taking place there, however independently of the quantity of people acting and living in this space. Thus is, that residential-only areas, which indicate usually relatively small numbers of inhabitants related to the surface, enjoy a higher protection than for instance mixed use areas, in which often substantially more humans live. These are suspended in numbers and in the height of traffic loads more strongly by harmful impacts, than humans in residential areas. It must be placed however surely, that in all spaces, in which a high number of humans stay during a long period of the day, a fundamental degree of health protection and quality of life is ensured.

Spaces for Public Needs

The FBC enables in accordance with § 5 (2) the representation of "spaces for public needs". With this space type however no statement about the sensitivity of the use herein ist connected. So in this space type, insensitive facilities e.g. sport halls and very sensitive schools or old people´s homes are summarized. In the future however should be considered about a necessary increased protection of sensitive uses which are located in these space types and be differentiated between "sensitive" and "lower sensitive" uses.

Special Areas – Other Types

In § 11 FUO are represented so-called "special areas – other types". By this, all types of otherwise not defined uses are summarized, also independent of their degree of sensitivity. Here the docklands e.g. can be called "very insensitive", whereas hospital areas must be classified as "very sensitive" in relation to traffic impacts. Also within the space type "special areas – other types" should be differentiated between "sensitive" and "nonsensitive" uses.

Green Spaces

The FUO offers the possibility for the classification of green spaces in accordance with § 5 (2). This space type however is usually neglected in the literature, although it possesses substantial functions within the fields recovery and leisure activities of the urban residential population. However, it is to be noted that also in this space type sensitive and insensitive uses are equally contained. Particularly rest-needy recreation sites such as cureparks, cemeteries, allotments and playgrounds are to be classified e.g. in relation to noise as "very sensitive". Sports grounds and public baths however serve "loud recovery" and should be kept as own space type.

Space Types

In the context of the research cooperative, the following space types were the basis for the assessment of traffic impacts:

1. Residential areas (residential-only, general, special)
2. Mixed use areas with above average proportion of residential population
3. Mixed use areas with below average proportion of residential population
4. Spaces for public needs with sensitive infrastructure facilities

5. Spaces for public needs with insensitive infrastructure facilities
6. Core areas
7. Commercial areas
8. Industrial areas
9. Sensitive green spaces for the calm recovery
10. Insensitive green areas for the active, loud recovery
11. Sensitive special areas
    - areas serving recreational purposes (§10 Federal Utilization Ordinance)
    - areas of tourism
    - spa areas
    - areas providing accomodation for visitors (§ 11 Federal Utilization Ordinance)
    - university areas
    - hospital areas
12. Insensitive special areas (§ 11 Federal Utilization Ordinance).

Village areas were not taken into consideration in this list. Since major roads attach in general to both sides to different area types and normally limit spatial uses, 78 possible street space constellations result from the space types mentioned. The classification of the street spaces must be done however only once with the build-up of the digital net model. Usually something will thus change only in the case of structural measures, i.e. within long periods. For these spatial constellations, sensitivity can now be defined in relation to traffic impacts by means of a four-level ordinal scale. Here it is to be considered however, that to the different space combinations no linear, absolute measure of sensitivity can be assigned. The degree of sensitivity can only be defined in connection with the type of traffic impacts, i.e. a criterion. As result of this consideration criterion-related limit values for maximum compatible traffic loads in built-up areas were assigned to the space constellations.

3 Criteria for the Description and Assessment of Traffic Impacts

As a next step, criteria for the description and assessment of traffic impacts were defined. In the literature common criteria frameworks frequently assess the traffic impacts with so called "planning criteria". For the assessment of time soluted traffic volumes however "control criteria" are needed and must be developed. Planning criteria usually refer to static components like technical or structural-formative aspects of urban and street spaces as well as to determining average traffic loads in larger view periods. The control criteria however are able to assess the time-step dependent impacts of increasing and decreasing high resolved traffic volumes (measured or simulated).

The sensitivity classification of spatial types was made on base of the following "control criteria":

- Noise,
- Pollutants,
- Separation effects.

These criteria represent a proven standard in the traffic planning.

By the example noise, now the steps are to be pointed out for the development of a control criterion for the description and assessment of the time soluted traffic volume.

The most important legal regulations and sets of rules in Germany, which contain limit values for the determination of the compatibility of noise, are:

- DIN 18005 "Noise Control in the Urban Development",
- 16. Federal Immission Control Regulation (traffic noise protection regulation),
- Guideline "Noise Protection at Federal Highways in the Building Load of the Federation".

As a result of the analysis can be noted, that the assessment of compatibility of traffic noise immissions is not constantly possible, because there are different

- areas of validity for the respective values,
- legal binding characters of the values (limit-, or orientation values),
- procedure levels (precaution level, redevelopment level),
- meaning levels (planning, system permission, new building of streets).

Furthermore, not for all urban space types exist limit values and there are no noise thresholds for existing streets underneath the redevelopment level.

Making it more difficult, the traffic noise calculation usually is done on base of the "Guidelines for the Noise Protection at Roads" (RLS-90) [2], with the reference calculation time "one hour".

So it was necessary, to define first a procedure for the time soluted calculation of noise immissions (time-step ½ h) and after that time dependent noise thresholds for all urban space types.

Calculation of Time Soluted Noise Immissions

The 16. Federal Immission Control Regulation says, that the evaluation level of traffic noise must be calculated in accordance with the remarks of the "Guidelines for the Noise Protection at Roads". Most important input for the calculation of noise immissions herein is the "decisive traffic volume". This is the average volume (all days of the year) of motor vehicles within the evaluation period (night/day), that pass the cross section of a road in one hour. The average daily traffic volume e.g. for the evaluation period "day" is broken down on one hour by multiplying it with 0.06. Result of the calculation is the average noise immission of one hour referred to the traffic volume of one year.

The calculation of the higher resolved noise immissions for the period ½ h cannot be made thus so easily on base of the "Guidelines for the Noise Protection at Roads". For a correct calculation of the noise level, the halfhourly simulated traffic volume is to be doubled. The result of this calculation is then however directly valid for the period ½ h, because with a duplication of traffic volume and calculation period, on base of the following equation

$$10_{\lg}(\frac{1}{2} * (10^{0,1L_1} + 10^{0,1L_1})) = 10_{\lg}(10^{0,1L_1})$$

the same calculation result is obtained. To sum up it can be said, that

- for a correct noise calculation on base of the "Guidelines for the Noise Protection at Roads", the halfhourly simulated traffic volumes must always be doubled and
- for a half hour the same legal limit values are valid as for a whole hour.

This rule can be applied now similar even to smaller time steps.

Assessment of Time Soluted Noise Immissions

The result of the noise immission calculation has to be checked, whether limit values are exceeded. Here it is to be considered however, that with the "Guidelines for the Noise Protection at Roads" a correct noise level can be calculated, but as the most important size the mean "decisive traffic volume" enters. As a result of that, an average and not time step dependent noise level, based on the traffic volume of one year, is received. That level sums up the different noise levels of working-days, sun- and vacation days and indicates beyond that daily load peaks in the context of rush hours as well as daily periods with lower traffic volumes. The mean noise level contains thus a lot of different high noise levels. These might exceed the limit values in some time-steps, without exceeding the legal limit value in the total calculation, because load peaks can be compensated by lower noise levels (e.g. late evening or sundays). It might be said that the noise levels oscillate around the limit values. The 16. Federal Immission Control Regulation allows thus the temporary exceeding of limit values and does not demand that the indicated limit values must be kept at each point in time, as long as the mean level does not exceed these. Now, if the halfhourly calculated noise level exceeds limit values, this does not mean at all, that the mean level is exceeded too. The answer of the question, whether time soluted noise levels exceed limit values and the following assessment of the noise, seems not to be possible on base of valid legislation.

For the solution of this problem, the relative (comparative) noise assessment was developed. Here the time-step dependent noise level is set in relation to the halfhour noise levels calculated so far in the evaluation period, i.e. for all levels of a complete evaluation period (day or night) an energetic mean value noise calculation is executed. Only if this mean value exceeds the thresholds, the noise level is no longer classified as compatible. By this methodology, brief threshold exceedings can be handled moderate in the noise evaluation. In order to receive a comparable mean noise pollution, the same number of noise levels must always be present, i.e. for the evaluation period "day" 32, and for the evaluation period "night" 16 halfhour values.

Determination of Limit Values for Time Soluted Noise Immissions

During the assessment of possible limit value exceedings on the basis of the relative noise level calculation, it is to be considered, that in one evaluation period the given limit values also may be exceeded in the ½ h intervals. To exceed limit values within the evaluation period is quite possible, without exceeding the limit value alltogether, if at other times lower levels are present. Thus the question comes up, which limit values of a relative noise assessment should be put as the base, because the 16. Federal Immission Control Regulation refers to the mean pollution of all days in the year. If the static limit values herein are taken, this has as a consequence that limit values at working-days are probably regularly exceeded, since lower noise levels at sun- /or vacation days, that could reduce the level, are not included into the calculation. The model description of the noise pollution in cities becomes thereby undifferentiated and little realistic. On the other hand the limit values in the 16. Federal Immission Control Regulation can and shall not be easily ignored, but also be the base of the further work.

To solve this problem, split limit values for noise pollution are suggested on base of the provisional version of the "Recommendations for Studies to Describe the Economic Efficiency of Roads" [3]. Examplarily, this split calculation shall be demonstrated for residential areas with a limit value of 59 dB (A) (day). The thresholds for a maximum compatible noise level, were developed by LUIS.

$$10\lg\left(\frac{1}{5.840}*(3.216(WD)*10^{6,0}+1.616(VD)*10^{5,8}+1.008(SD)*10^{5,7})\right)=59dB(A)$$

5840 = number of hours for the evaluation period "day" in one year

For working-days (WD) a limit value of 60 dB (A), for vacation working-days (VD) 58 dB (A) and for sundays (SD) 57 dB (A) is set. The limit values for vacation working-days and sundays result inevitably from the raised limit value for working-days. Higher limit values cannot be assumed for the latter two daily types, since hereby the total limit value of 59 dB (A) is exceeded. This also means, that the limit value for working-days cannot be higher, since otherwise the limit values for sun- and vacation working days must be set unrealistically low. The limit values for the evaluation period night should not be split, since for this hardly drive-purpose-dependent modifications in the traffic loads can be proven.

Definition of Spatial and Time Soluted Limit Values

From the preceding remarks, the following split noise limit values for the specific space types can be derived:

Area Designation	Sensitivity Criterion: Noise	Limit Value				
		Day				Night
		total	WD	VD	SD	
Residential areas	2	59	60	58	57	49
Mixed use areas 1	2	59	60	58	57	49
Mixed use areas 2	3	64	65	63	62	54
Spaces for public needs 1	2	59	60	58	57	49
Spaces for public needs 2	3	64	65	63	62	54
Core areas	3	64	65	63	62	54
Commercial areas	4	69	70	68	67	59
Industrial areas	4	69	70	68	67	59
Green spaces 1	2	59	60	58	57	49
Green spaces 2	3	64	65	63	62	54
Special areas 1	1	57	58	56	55	47
Special areas 2	4	69	70	68	67	59

Table 1: Split noise limit values for different space types.

1 = very high sensitivity

2 = high sensitivity

3 = middle sensitivity

4 = low sensitivity

When defining limit values for street space combinations, the higher sensitivity of a roadside space is always taken.

With the other "control criteria", needed for the assessment, a similar methodology takes place. First, the degree of the spatial sensitivity in relation to the respective criterion must be determined on base of a four-level ordinal scale. After that, the limit values can be defined.

Space Types	Description
Mixed use areas 1	Mixed use areas with above average proportion of residential population
Mixed use areas 2	Mixed use areas with below average proportion of residential population
Spaces for public needs 1	Spaces with sensitive infrastructure facility
Spaces for public needs 2	Spaces with insensitive infrastructure facility
Green spaces 1	Spaces for the calm recovery
Green spaces 2	Spaces for the active recovery
Special areas 1	Special areas with sensitive infrastructure facility
Special areas 2	Special areas with insensitive infrastructure facility

Table 2: Description of the space types.

In the next step, the simulated traffic volumes can be linked with the sensitivity of the affected spaces to assess the compatibility of the traffic load. This takes place on base of the computer-assisted information system LUISE.

4 The Information System LUISE

The information system LUISE is able:

- to display the maximum compatible traffic volume in street-spaces (based on the sensitivity of spatial uses) and the simulated traffic volumes for each section of the street net (main roads).
- to visualize the changes of time soluted traffic volumes.
- to compare the actual value with the set-value (limit value) and to make calculations for the determination of compatibility.
- to display the exceeding of thresholds and the degree for all criteria.

Fundamental parts of this system are the digital main street net and the digital preparatory land-use plan for the city of Wuppertal, which were integrated into the geographic information system Arc-View and completed with user-specific informations and programmings.

The data bases belonging to the street net contain each necessary information for the description and assessment of the traffic volume on base of the criteria framework for 48 half hours of the day, for all street sections. Thus, each segment contains the attributes, which street space constellation and degree of sensitivity in relation to the defined criteria is given, as well as the maximum compatible traffic volume (number of vehicles/½ h) can be for each criterion.

In LUISE the net-wide traffic volume for every half-hour of the day can be entered and be assessed referring to a selected criterion. For this, system internal a comparison between the momentary, i.e. the selected halfhour actual volume of the section, with the maximum compatible set-value takes place (Fig.1).

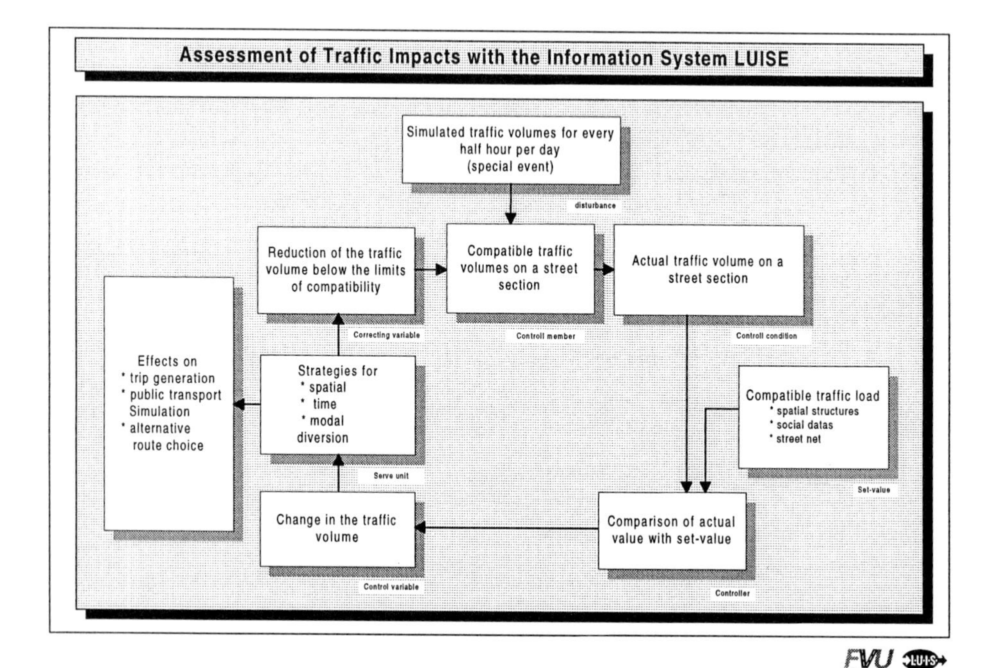

Figure 1: Control loop mechanism for the assessment of traffic impacts in the information system LUISE.

For the assessment of noise pollution this means e.g., that first system internal noise immission calculation of the actual traffic load is executed. The result of the calculation is compared with the set-value (limit value) and displayed for every street section on base of a three-level ordinal scale (no limit value exceeding - moderate limit value exceeding - high limit value exceeding) with different colours. If the number of vehicles of the actual load exceeds the defined limit value, this status is optical concisely emphasized, so that it is opened for the viewer immediately, whether measures for traffic diversion must be initiated (Fig. 2).

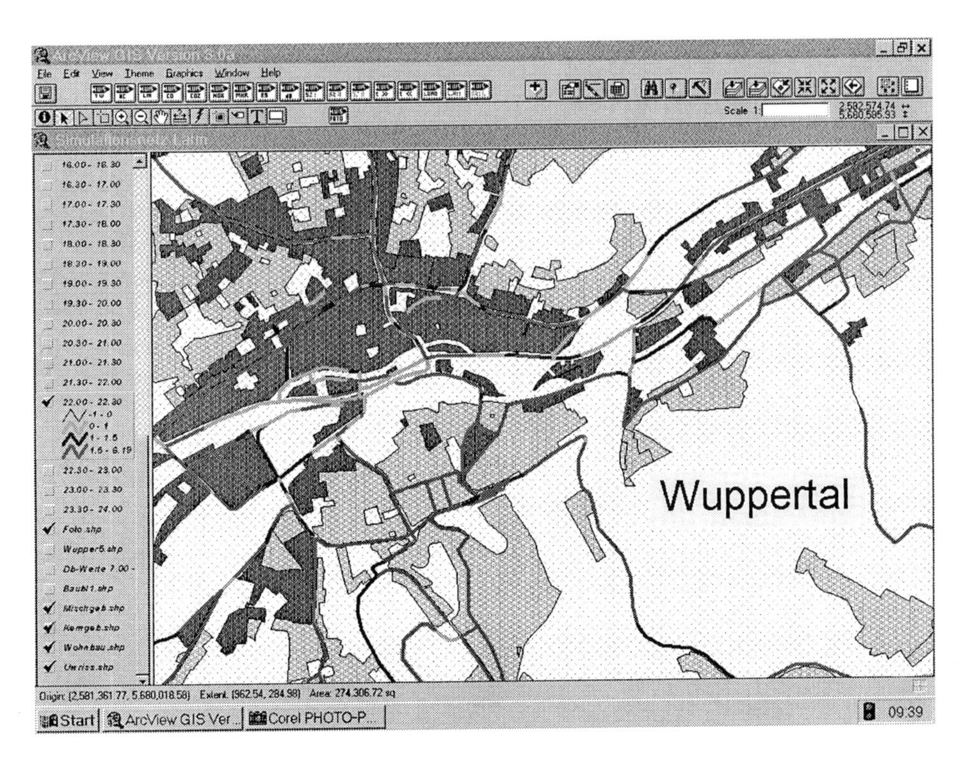

Figure 2: Display and assessment of traffic noise in the area Wuppertal-Elberfeld with the information system LUISE.

The border represented within the left area of the figure shows the time period, for which the assessment was made (10.00 p.m. – 10:30 p.m) and the degree of limit value exceeding. In the upper area of the figure, criterion-related "buttons" are specified, with which the impact situation for the criterion concerned is represented. For the better visualization of the street space situation, additionally photos can be loaded from selected street sections. By the assessment and display of traffic loads in cities on base of LUISE and the passing of these data to the traffic control, here e.g. to dynamic navigation systems, traffic load statuses of the superordinate and the subordinate network can be linked together. Thereby, compatible route recommendations for both net types can be determined.

The system LUISE is portable to all cities, if spatial uses and the road system are digitally present. It should be led for good reason at regional traffic control centers, since all information about traffic load statuses in the superordinate and in the future also the subordinate net, gathers here. With such a global knowledge of the actual traffic status, recommendations for a compatible route selection can be expressed and transmitted over mobile phone or RDS-TMC into the vehicles.

5 Summary

The part of LUIS within the research cooperative was to assess the time soluted impacts of traffic in towns. It can be noted, that common criteria frameworks do not fit to this demand. New criterias, especially for the traffic control had to be developed. Regarding to the space types in towns it can be said, that these are mostly not classified referring to their sensitivity to traffic impacts. New space types in towns had to be determined. The guidelines for the calculation of noise can not be easily transfered on the time-step ½ h. Further calculations must be done to get a realistic noise level. With LUISE, a computer based information system was created, that is able to calculate outgoing traffic impacts and to compare this actual value with a set-value for every half an hour of the day. With such a knowledge of the actual traffic status in towns, recommendations for a compatible route selection can be expressed by dynamic navigation systems.

References

1. Forschungsgesellschaft für Straßen- und Verkehrswesen (Hrsg.) *Richtlinien für die Anlage von Straßen, Teil: Leitfaden für die funktionale Gliederung des Straßennetzes (RAS-N),* Köln (1988).

2. Forschungsgesellschaft für Straßen- und Verkehrswesen (Hrsg.) *Richtlinien für den Lärmschutz an Straßen (RLS-90)*, Köln (1990).

3. Forschungsgesellschaft für Straßen- und Verkehrswesen (Hrsg.), *Empfehlungen für Wirtschaftlichkeitsuntersuchungen an Straßen*, nicht veröffentlichte Fassung, (1997).

4. Y. Zhao, *Vehicle Location and Navigation Systems* (Boston, Artech House, 1997).

5. Senatsverwaltung für Stadtentwicklung und Umweltschutz (Hrsg.), *Studie zur ökologischen und stadtverträglichen Belastbarkeit der Berliner Innenstadt durch den Kfz-Verkehr*, in: Arbeitshefte Umweltverträglicher Stadtverkehr, Heft 4, Berlin (1993).

6. P. Müller, J. Collin et al., *Das LADIR Verfahren zur Bestimmung stadtverträglicher Belastungen durch Autoverkehr*, Schlußbericht zum Forschungsprojekt des Forschungsfeldes „Städtebau und Verkehr“ im Bundesministerium für Raumordnung, Bauwesen und Städtebau, Darmstadt/Braunschweig (1994).

7. G. Siegle (Hrsg.), *Telematik im Verkehr*, R. v. Decker´s Verlag, Heidelberg, (1996).

Lecture Notes in Computational Science and Engineering

Vol. 1 D. Funaro, *Spectral Elements for Transport-Dominated Equations.* 1997. X, 211 pp. Softcover. ISBN 3-540-62649-2

Vol. 2 H. P. Langtangen, *Computational Partial Differential Equations.* Numerical Methods and Diffpack Programming. 1999. XXIII, 682 pp. Hardcover. ISBN 3-540-65274-4

Vol. 3 W. Hackbusch, G. Wittum (eds.), *Multigrid Methods V.* Proceedings of the Fifth European Multigrid Conference held in Stuttgart, Germany, October 1-4, 1996. 1998. VIII, 334 pp. Softcover. ISBN 3-540-63133-X

Vol. 4 P. Deuflhard, J. Hermans, B. Leimkuhler, A. E. Mark, S. Reich, R. D. Skeel (eds.), *Computational Molecular Dynamics: Challenges, Methods, Ideas.* Proceedings of the 2nd International Symposium on Algorithms for Macromolecular Modelling, Berlin, May 21-24, 1997. 1998. XI, 489 pp. Softcover. ISBN 3-540-63242-5

Vol. 5 D. Kröner, M. Ohlberger, C. Rohde (eds.), *An Introduction to Recent Developments in Theory and Numerics for Conservation Laws.* Proceedings of the International School on Theory and Numerics for Conservation Laws, Freiburg / Littenweiler, October 20-24, 1997. 1998. VII, 285 pp. Softcover. ISBN 3-540-65081-4

Vol. 6 S. Turek, *Efficient Solvers for Incompressible Flow Problems.* An Algorithmic and Computational Approach. 1999. XVII, 352 pp, with CD-ROM. Hardcover. ISBN 3-540-65433-X

Vol. 7 R. von Schwerin, *Multi Body System SIMulation.* Numerical Methods, Algorithms, and Software. 1999. XX, 338 pp. Softcover. ISBN 3-540-65662-6

Vol. 8 H.-J. Bungartz, F. Durst, C. Zenger (eds.), *High Performance Scientific and Engineering Computing.* Proceedings of the International FORTWIHR Conference on HPSEC, Munich, March 16-18, 1998. 1999. X, 471 pp. Softcover. 3-540-65730-4

Vol. 9 T. J. Barth, H. Deconinck (eds.), *High-Order Methods for Computational Physics.* 1999. VII, 582 pp. Hardcover. 3-540-65893-9

For further information on these books please have a look at our mathematics catalogue at the following URL: `http://www.springer.de/math/index.html`

Druck: Strauss Offsetdruck, Mörlenbach
Verarbeitung: Schäffer, Grünstadt